HOLODYNAMICS

HOLODYNAMICS

*How to Develop and Manage
Your Personal Power*

V. Vernon Woolf, Ph.D.

Foreword by
F. David Peat

∞This book was printed on acid-free, archival-quality paper
Typeset in Palatino 10/14 on a Macintosh SE
Designed by Kathleen Koopman

Library of Congress Cataloging-in-Publication Data

Woolf, V. Vernon. 1937-
    Holodynamics: how to develop and manage your personal power/ by
V. Vernon Woolf.
        p.   cm.

    1. Self-actualization (Psychology) I. Title
BF637.S4W66 1989
158'.9--dc20   89-39181
                                                        CIP

Contents

Foreword

Vernon Woolf's *Holodynamics* is an exploration of the blocks and rigidities that restrict a person's life and prevent him or her from entering fully and creatively into relationships with friends, family, society, life work, and, indeed, the whole world. People become imprisoned by these patterns of behavior response; but they are bound by chains of their own making, Woolf argues. The central issue of his book is: how can people free themselves?

As a physicist, I cannot endorse or comment upon the therapeutic validity of Woolf's work—this lies beyond my area of expertise. But I do believe that physics has been responsible for reinforcing a certain way of looking at the world that has driven a wedge between our thinking and feeling, between our minds and our bodies, between ourselves and society, and between society and the whole planet. This worldview, which extends far beyond science, has eroded the meaning from our lives and left us with a feeling of emptiness and powerlessness.

Over two hundred years ago, Newton, standing on the shoulders of the giants who had come before him, created a vast and comprehensive theory of the universe. But Newton's vision was essentially mechanical in nature: everything that existed was built out of smaller elements which were in interaction with each other, everything that happened had a mechanical cause, and every cause had a predictable and deterministic outcome. The universe was a giant clockwork—and seemed to have no room for men and women with their dreams and aspirations.

This mechanistic vision became the paradigm for the other sciences which aspired to the rigor of physics. Indeed, Newton's vision of the world spread even further: the philosopher and political scientist John Locke described himself as "an under-gardener" working in the shadow of "the incomparable Mr. Newton." Even Sigmund Freud attempted to base his theories of psychoanalysis upon a strictly scientific model—psychopathology was to be seen in terms of the flows, checks, balances, blocks, and repression of energy.

What had begun as a purely scientific description of the motion of planets and cannon balls was now having a very profound effect upon our lives. It influenced the very way we saw ourselves. It dehumanized the universe by placing us outside, as objective observers whose goal was to predict, control, and dominate nature. From this objective position, observing a dead, mechanical world, we could no longer enter into the living essence of things, we were exiled from our own bodies and even from our feelings. Believing that change could only come about by mechanical causation and that separation could only be bridged by force, we became orphaned from the universe. But Vernon Woolf and others are now suggesting that this period of isolation is ending and that we now see ourselves as empowered to change our lives and enter fully into the future of our planet.

The break in this logjam of thought occurred in the first decades of this century, when the Newtonian worldview was split apart by the new insights of quantum theory. But these insights have yet to fully percolate out of physics and transform our common way of seeing nature and ourselves. Readers may have heard that God plays dice with the world, that an electron has a curious wave and particle duality, or of Heisenberg's Uncertainty Principle. But they may be less aware of the central, revolutionary premise of quantum theory—that in any perception observer and observed are one and unanalyzable. Quantum theory has brought a new holism to our view of nature—men and women no longer stand outside the universe as isolated observers but have been embraced into the heart of nature as full participators. To extend this metaphor, mind and body have become one, thinking and feeling are no longer fragmented. No longer is the universe deterministic and predictable. From now on, we shall work within nature in an entirely new way.

Even the very foundations of matter have begun to change. The physicist David Bohm speaks of the "holomovement"—a ground of flow out of which everything unfolds. Today matter is no longer seen as hard and fast but as patterns in a flux. The elements of this new reality are like a vortex that forms in a river. This vortex may have the appearance of an independent entity, yet it owes its birth, death, and continued existence to the flowing river. Woolf's "holodynes" are images of this vortex. No longer can the patterns of our lives be viewed as fixed and rigid—they are all manifestations of an underlying creative flux and for this reason are themselves subject to change.

The new visions suggest that matter and mind are no longer separate; in fact, they may be different aspects of a single, unifying process. Just as matter is constantly forming and dissolving, so, too, thought and mind are in a constant process of birth and death. But just as certain material forms tend to persist and become the "habits" and inertia of the material world, so, too, within the flowing movement of thought, certain forms crystallize and rigidify. The problem is that these forms fail to change with the world around them. They continue to evoke habitual responses and fixed behavior patterns that are inappropriate or fragmentary when taken in the context of an evolving family life or work situation. Ultimately such rigidities, or "holodynes," as Woolf calls them, stretch out into social and political life, giving rise to the tension, conflict, and violence that we see in the world around us.

Woolf's new vision suggests that we are not inevitably stuck within these holodynes. They are all part of a much deeper flow; so that if our lives have become dominated by them, it is because we are actually giving our energy up to them—we are constantly forging the chains that bind us. But change *is* possible. We can die to the past. *Holodynamics* replaces a rigid, deterministic worldview with one that is vital and creative.

F. David Peat

Acknowledgments

This book is the distillation of two decades of research and practical experience. The principles and processes have been refined through the years when I worked as a therapist, and to all my patients I give special thanks. To all those who have put this information to work in different community and private programs, and to those thousands who are using it in their private lives and who are part of the network of unprecedented support that surrounds this book, I express my heartfelt loving thanks. I would like to pay a special tribute to:

My editor, Jeffrey Lockridge, whose depth of perception and breadth of comprehension helped refine and condense these findings, and Laurel Gregory and her team at Harbinger House who have provided the detailed professional service necessary for the publication of the book.

My family whose support has allowed me space and time to write and whose love has made the birth of the science of holodynamics possible, and to my mother, Lena Ross, who laid an enlightened foundation and set the example for my path in life.

My friends who organize the conferences across the country so these principles and processes can be taught in various disciplines and to people everywhere.

Those scholars from various disciplines who have helped me understand and explain holodynamics; the list is almost endless, but I would like to mention just a few who helped my thinking in quantum leaps: from quantum physics, David Bohm, F. David Peat, Ken Wilber, and Fred Alan Wolf; from quantum chemistry, Ilya Prigogine, Rupert Sheldrake, and Lewis Thomas; from neurophysiology, Karl Pribram (holographics), Cleve Backster, Robert Becker, Bjorn Nordenstrom, Gary Selden, and Roger Sperry (resonating frequencies of the body).

Murray Bowen, John Bradshaw, James Framo, William Glasser, Virginia Satir, Carl Whitaker, and a host of my friends in the area of marriage and family therapy.

Patricia Cota-Robles, Gerald Jampolsky, Sam Keen, Ken Keyes, M. Scott Peck, Jacquelyn Small, Ron Smotherman, Hal Stone, and Sidra Winkelman, and all those pioneers of human thought who are shifting the field of human consciousness.

Stephen R. Covey, Paul H. Dunn, Neil A. Maxwell, Carol Lynn Pearson, and those who continually inspire my personal life.

My colleagues and friends who have waited so long for this book, spent endless hours discussing various aspects of the scope of this approach to life, and all those who will now apply what is put forth in this book.

HOLODYNAMICS

Introduction

I would like to present to you eight of the most powerful principles in the universe.

These principles are practical. They have been tested again and again. They work. They work for individuals, for families, for businesses. They work on the street. They help solve some of the deepest problems facing mankind. And because they work, they hold out hope for the future.

They were not always apparent to me as I was called late at night to work with people in crisis, or knelt beside an addict or an alcoholic in the street, or counseled parents of adolescents in trouble, or helped someone release deep anger in therapy. But gradually, over the years, I came to understand that something very powerful is at work in every one of us, and that each life, no matter how tragic or triumphant, unfolds according to these same basic principles.

The Eight Principles of Holodynamics

1. The universe is holodynamic. All matter, energy, and intelligence—past, present, and future—are part of one, dynamic whole.

2. The universe contains living thought-forms, called "holodynes," with the power to manifest reality in all dimensions.

3. The universe has an underlying, enfolded order—the Implicate Order.

4. Within the Implicate Order, the mind, holodynes, human beings, and all manifest reality follow the *same* six stages of development.

5. Manifest reality has both a "particle" and a "wave" function, which the mind reflects through its rational and intuitive processes, respectively.

6. Change occurs holodynamically: to change any holodyne is to change the physics of the mind; to change the physics of the mind is to change the physics of the universe—past, present, and future.

7. Every human being has a primary, controlling holodyne, called the "I" or "Full Potential Self."

8. The holodynamics of the mind can be applied systematically to solve every problem of human experience.

As I sensitized myself to what really worked for people, how they "came into" themselves, how they aligned themselves with health and growth, how they solved their problems, and then how relationships—and even systems—matured and solved *their* problems, I found that these principles were *universal*: they applied at *all* levels of human experience.

To understand holodynamics, I needed to go beyond the limits of conventional thinking—beyond psychology, philosophy, religion, and their Newtonian worldview—to the new sciences of the Quantum Age.

My own experience had taught me that intuition was by far the greatest gift I could have as a therapist—that all the rational training I had gone through was of very little use in healing and helping on the street. Somehow it was intuition that really counted. But how could intuition be explained from a rational perspective? Sometimes I could sense intuitively what the problem was, even though I had little rational information about it. That deep sense began to fascinate me. I taught myself how to apply my sensory perception processes to the intuitive part of my mind. This opened doors that had never been opened before. Learning to use my senses on my intuitive mind was one of the greatest experiences of my life, and I found myself in possession of the greatest computer I could ever imagine. It was my intuitive mind—my right brain—that took in the totality of things from a "wave" perspective—that dealt with things as part of one, dynamic whole.

From quantum physics, I knew there was an underlying order throughout the universe and an all-pervading force, which manifested itself in the dual nature of reality: as *both* "particle" *and* "wave."

From holographics, I learned how the mind stored memories; from the developmentalists, how the mind developed according to definite stages; from the neurologists, how the mind had causal potency.

From fluid dynamics, I learned how "shape" within a "stream" influenced the whole stream; from topology, how shape could be changed without changing function; from chaos theory, how every chaotic system was part of a greater order and had within it its own hidden order.

Then everything came together.

I looked at the mind from a quantum perspective and began to understand how the mind had a direct impact upon reality.

I looked for "shapes" within the "streams" of thought and so discovered "holodynes," and then, the "Full Potential Self"— the "I."

I needed a map of the mind and a way to apply it to the dynamics of human experience, and thus evolved the "Mind Model" and "phase-spacing."

I needed a process to help mature the primitive holodynes within the mind and another to help someone reach his or her fullest potential in the outside world, and thus evolved "tracking" and "potentializing."

And I needed to be clear about how everything worked within the order of things, and thus evolved the "principles of holodynamics."

At the very heart of these principles, at the most fundamental level of personal being, is a force I call the "I." It's in every one of us. It is you at your fullest potential—your "Full Potential Self"— the seed within you that unfolds, grows, and blossoms through all of your experience.

The "I" cannot be potentialized in isolation. It is holodynamic: *both* individual *and* universal—inseparably part of the one, dynamic whole.

Through the processes of tracking and potentializing, you can align yourself with the "I" so that your life unfolds, naturally and easily, to its fullest potential. These processes, adjusted and fine-tuned over the years, proved invaluable from the very start. I have taught them to thousands of people, to therapists, doctors, educators, and businessmen all across the country, and have taught others how to teach them.

Here, for the first time, I have explained tracking and potentializing and the principles behind them so that everyone can learn how to use them.

The Mind Model and phase-spacing will help you solve "unsolvable" problems. You will discover, through holodynamics, how you can *create* reality—in harmony with yourself, with others, and with the universe.

1

Basics

The Power of "I"

To begin the study of holodynamics (from "holo," meaning "whole," and "dynamic," meaning "force in action"), I want to focus first upon the "I," the Full Potential Self—the primary unit of personal being and power in the holodynamic universe, and the pivotal point of all human experience. It is your "I" which creates all of the circumstances that fill your life, control your thinking, and determine your success. Within your "I" are contained all the keys to your past, present, and future. Your "I" is the seed from which you spring and the full potential toward which you strive. Your "I" is your "completeness" enfolded within the fabric of your nature, and whether it unfolds or not is entirely up to you.

You, the conscious, aware self, get to choose whether your "I" unfolds or remains forever enfolded. Every moment of every day you have been choosing, not consciously perhaps, but choosing nevertheless. Every time you focus, think, feel, or act, you choose. Every word you speak reflects the choices you make—sometimes at very deep levels and sometimes superficially—choices which take place in the orders within orders of your mind. Your entire life has been especially designed for the unfolding of the "I" within you. All that you feel and everything that you wish for is a reflection of your "I."

Focus, for example, on what you really want. Do you want peace? Do you want more money? Do you want a deeper, more intimate relationship? Do you want more fun in your life? Do you want to have greater success? Better health? More happiness? Do you want to be a better friend? A better parent? A better leader?

How then, do you get from where you are to where you want to be? How do you get from point A to point B?

First, you must understand that you are exactly where you have chosen to be. You are making exactly the amount of money you have chosen to make. You have exactly the kind of friends you have chosen to have. You are exactly the kind of parent you have chosen to be. For what you experience in life are the consequences of choices made deep within your mind, often subconsciously, but made in a part of your mind which is intelligent, knowing, and caring. Even your most miserable, painful, and devastating experiences are the result of choices you have made. Not the conscious you, but the "I" within you. This "I" controls all the orders within orders within your mind.

To align your conscious self with your "I," it is necessary to access the deeper dimensions of your mind and to learn the principles by which everything operates. You will learn that your mind is a reflection of the total universe, that it is, in fact, *holodynamic*.

There is perhaps no better reflection of the orders of nature—the natural, underlying, enfolded orders within orders—than the human mind. It is the clearest, purest, and most accessible expression of life on this planet—and also the most complex. As such, the mind gives us our most comprehensive view of nature's orders within orders.

Orders within Orders

Everything you have ever believed is not true.

It's not that it's false, it's just not true in the way you believe it's true. For even the way you look at things, the way you interpret them and feel about them, is subject to orders within orders.

It reminds me of the story told of the scientist who was giving this wonderful lecture on how the world was formed, and how the sun is the center of our particular solar system, and how everything is held in natural balance. After the lecture a little old lady came up to him and said, "You can't fool me, sonny. This earth is flat and it sits on the back of a great turtle." The scientist smiled and in a condescending way said to her, "Well, madam, if this earth is flat and sits on the back of a great turtle, upon what does the turtle sit?" "Don't get smart with me, sonny," she replied indignantly. "It's turtles, all the way down!"

In a way, the lady may have been right. Take, for example, the beautiful patterns within patterns which Benoit Mandelbrot discovered in the random number streams of simple formulas *(see Plate I)*. They really do look like "turtles all the way down." Such patterns are everywhere in nature—in leaves, in snowflakes, in river courses—and everywhere they flow from underlying orders— orders within orders—which are themselves mathematical. In fact, from simple formulas for such orders, computers can generate every known pattern of nature, no matter how intricate or complex. Somewhere in every natural pattern,"it's turtles all the way down."

To discover the orders within orders hidden within random

streams of numbers, Mandelbrot "phase-spaced" them. That is, he looked at the numbers from a completely different "space" and explored their "phase" of development. Using this same process, you can "phase-space" the streams of thought within your mind and discover the orders within orders hidden there.

Phase-spacing

What you see is not all there is.

To explore the orders within orders of your mind, you phase-space your mental patterns. You take the natural way you look at things and put it into another space, so that you can see your patterns from a different angle. For example, suppose you were asked to draw the motion of a pendulum on a piece of paper. How would you draw it? Now if you phase-spaced it—saw it from a different angle—the motion of a pendulum might look like the figure below:

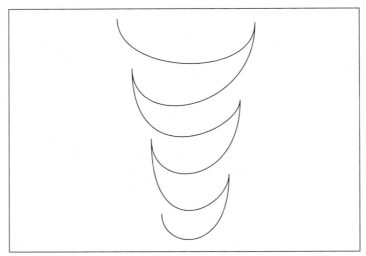

FIGURE 1

It's like being *in* a ball of string. If you're totally wound up in the ball of string, your whole world *is* the problem, or the ball of string. But if you step *out* of the problem, *out* of the ball of string, so you can see what it is, you can get a much better view of how to approach whatever it is you have there, whether to untangle the ball of string, or simply to walk away from it—until it becomes just a dot in your road of life. Chances are, you'll have to deal with your problem, sooner or later. And when you do, phase-spacing will be there to help you.

And so it is that you can phase-space all kinds of problems, just as you can mathematical formulas. The ability to phase-space has helped catapult science forward—into the Quantum Age—to discover as much in the last two years alone as in all of recorded history up until then. And phase-spacing, by letting you see and understand the hidden orders within your mind, can help you use this new knowledge with great power.

If you are in depression, for example, you live in blackness. It's as though you are wrapped up within it. But as long as you are inside of the blackness, you cannot see your way out of it. So you need to do what anyone wrapped up in a problem needs to do, first and foremost—find a way to step outside the problem and see how it functions for you. If your mind is in chaos, you can borrow a little bit from chaos theory, look at the chaos of your mind, and do what scientists do when they want to find out about chaos anywhere in nature—*phase-space* it. The same process that helps them understand the orders within orders of chaos will help you understand the orders within orders of your mind.

Therefore, if you are wrapped up in a problem like depression, your first job is to learn how to phase-space. For even though the whole world may look like blackness, there is a part of your mind that can independently step outside all that blackness and phase-space it and see it for what it really is.

You can apply phase-spacing to any problem. If you are in the middle of a temper tantrum, you can phase-space it. If you are not as successful as you would like to be in your life, you can phase-space success. You can look at the dynamics of a complete relationship or a complete system by phase-spacing it. And you can do the same thing with principles, and with the universe.

Once learned, phase-spacing becomes a primary tool for anyone wishing to understand the deeper orders of the mind. The process of phase-spacing gives you the freedom to deal with any kind of dynamic and to solve any kind of problem. When skillfully used, it provides you with all the tools you need to direct your mind and systematically unfold your fullest potential in the physical universe.

It was phase-spacing that showed me the power of the "I," the Full Potential Self working within every individual. For no matter what problem anyone had, whenever I phase-spaced it, I found a Full Potential Self orchestrating that problem. So in this book, I will teach you how to phase-space and give you a universal model and a number of workable, tested ways to solve any problem. You will find that the world looks different when you phase-space it— as different as a full-color Mandelbrot set from a page of random numbers. You will find that life is far better, far more beautiful than you ever dreamed it could be. And you will find that problems you thought would drown you are but ripples in the Quantum Wave and are all provided for your pleasure.

But if you want to understand the nature of your problems, and the orders within orders that are creating those problems, you must be prepared to accept that everything you have ever believed is not true. Don't count it as false, just realize that when you phase-space it you will see it in an entirely different way, and with that new insight you will learn to unfold the wonderful potential within every problem you have ever had. You will learn there never was a problem that didn't have built into it a lesson, and there never was a problem that your mind didn't know the solution to before it created the problem.

How then, do you shift from problem focus to solution focus? How do you align your life energies to become part of the solution to any problem? How do you really get from point A to point B?

You start by phase-spacing reality. If you want to solve any problem, you must begin at the beginning—with the very nature of reality—with the "particle" and the "wave." To phase-space reality, you step outside your normal view of things: you "un-mesmerize" or "de-hypnotize" yourself. You shift from your rational view to a quantum view and come to see reality through the eyes of a quantum thinker.

Quantum Thinking

Quantum thinking is some of the best thinking around right now. In fact, quantum thinkers have devised the most accurate science ever created by humankind—one that predicts better and explains more comprehensively than any other—everything from subatomic particles and biological systems to space travel and the movements of galaxies. Quantum thinking began around the turn of the century when scientists, using classical (Newtonian) physics, could not account for the stability of the atom or the amount of energy radiating from a "black body." A new perspective was needed, and thus was born quantum mechanics. Max Planck and Albert Einstein solved the riddle of black body radiation by giving "quantum" measurement to the radiation field, and thus was born the "photon." Niels Bohr, Louis de Broglie, and Erwin Schrödinger solved the riddle of atomic stability by giving "quantum" measurement to energy waves within the atom, and thus were born the "standing waves" of the electron.

But viewing radiation fields and energy waves as "quanta" (discrete "bits or pieces" of energy, which are *both* "particles" *and* "waves") sent ripples throughout the world of science. Such thinking implied that there might be dimensions of reality beyond the speed of light which were causing things to happen here. It suggested answers to otherwise puzzling questions. What, for example, was making the very fabric of space bend? And why was there an irreducible "uncertainty" to everything in the universe?

When scientists looked at the universe from a quantum perspective, they discovered in the very fabric of space and time an all-pervading force—the Quantum Force—so great that a cubic centimeter of space contained more energy than all matter everywhere. They discovered that this force spread throughout the universe in an infinite "field"—the Quantum Wave. And they discovered that the Quantum Force, the Quantum Wave, and everything in the physical universe were governed by an underlying order, an "order within all orders."

Three central concepts of quantum thinking helped me explain what I was observing about the human mind. First was the concept of a universal field containing all possibilities, every potential for any circumstance: the Quantum Wave. Second was the concept of an underlying, enfolded order throughout the universe: the Implicate Order. And third was the concept of the dual nature of physical reality: *both* "particle" *and* "wave."

THE QUANTUM WAVE

The Quantum Wave was thought, at first, to be only a theoretical probability, but David Bohm and others, who place it in what they call the "unmanifest plane," are convinced from their findings that the Quantum Wave is the source of all physical reality—of everything in the "manifest" plane. So everything we touch, see, hear, smell, and taste, everything in our physical world, is the result of a "collapse" of the Quantum Wave into the "particles" of physical reality.

Some scientists call the Quantum Wave the "Offer Wave" because, in a way, it "offers" every possibility. The Offer Wave is seen, from a quantum perspective, as being composed of an infinite number of "quanta" ("bits and pieces"), which make up the complete set of possibilities for any given circumstance. It is the human mind which triggers the "collapse" of the "wave" function. When you *choose* what you want, you focus on one particular "quantum." By *choosing* it—and meeting all the boundary conditions, you cause the Quantum Wave to collapse, and what you want becomes a reality.

THE IMPLICATE ORDER

David Bohm, one of the most gifted quantum thinkers of the last fifty years, hypothesizes that all matter, energy, and even time itself are the result of an order enfolded within the universe: the Implicate Order. This deep order is contained in what he calls the "quantum dimension," or within the Quantum Wave itself. The Implicate Order governs everything, living and non-living, from the smallest cells and subatomic particles to human societies and the universe itself. It explains why species migrate, why things grow in all the complex patterns of nature, how life organizes itself, and most importantly, how the human mind functions—how it *creates* reality.

For Bohm, the Implicate Order is "non-linear": he sees the universe as a "field of influence," in which all material things and all the laws of nature reside—as one, dynamic, interacting quantum projection. I call this universe the "holodynamic universe."

THE "PARTICLE" AND THE "WAVE"

From a holodynamic perspective, everything is made of energy. The Quantum Force "manifests" this energy in the physical world as both "particle" and "wave." This point was made by Albert Einstein, Boris Podolsky, and Nathan Rosen (E.P.R.) over fifty years ago. They proposed that when any two coupled particles were separated and one was measured as either a "particle" or a "wave," it would "collapse the wave function"for the other to become the same thing, no matter how far apart they were. In other words, measuring one to be a "particle" or a "wave" would make the other a "particle" or a "wave." In 1965, John Stewart Bell devised a way to actually perform the E.P.R. experiment.

In Bell's experiment, coupled photons were split apart, and as the photons traveled away from one another, each was measured to see if it was a "wave" or a "particle." If one photon was measured to be a "wave," sure enough, the other photon also turned out to be a "wave." But if one photon was measured to be a "particle," then the other photon "became" a "particle." So there was some way that one photon was explaining to the other, "Hey, they've measured *me* to be a particle, so *you* be a particle." And to do that they had to communicate faster than the speed of light. Because the two photons were traveling away from one another, each at almost the speed of light—and *together* at *almost twice* the speed of light. Even so, what Einstein, Podolsky, and Rosen proposed, and Bell demonstrated, turned out to be true: there *was* a dimension beyond the speed of light and it *affected physical reality.*

The E.P.R.-Bell experiment also showed that your results depend in part on your measuring apparatus, and in part on what you *intend* to get. Whether something appears as a "particle" or a "wave" depends on how you measure it—how you look at it—and on what you *want* it to be. Elementary "particles" or "waves" behave according to the conditions *you* set for them, and so, therefore, does physical reality.

If you step outside of your usual thinking processes for just a moment and phase-space what is happening in your view of reality, one thing becomes evident—your mind has two major processes by which it deals with reality—a rational ("particle") process and an intuitive ("wave") process. You can look at anything from a "particle" perspective or a "wave" perspective, and reality will usually respond.

Quantum thinking, then, gives us a powerful, new perspective not just on physical reality, but also on the human mind as it perceives, affects, and *creates* that reality. Quantum thinking helps us understand and direct our minds better than any other science ever devised.

Topology

One way to keep track of phase-spacing the dynamics of reality and the mind—the interactions of the rational and intuitive processes with the Quantum Wave, the Quantum Force, the Implicate Order, and the "particle" and "wave" functions of reality—is to create a topology. Scientists and engineers use topology to change the size or shape of something without changing any of its functions. So, by metaphorically adopting topology, we can "map" the mind in the shape of a simple cube, without changing any of its functions. Now, in light of the new quantum sciences, let's look at a topology of the mind, a mind model which shows both the "particle" and "wave" functions of reality and which takes into account all other aspects of the quantum view of reality.

Plate I

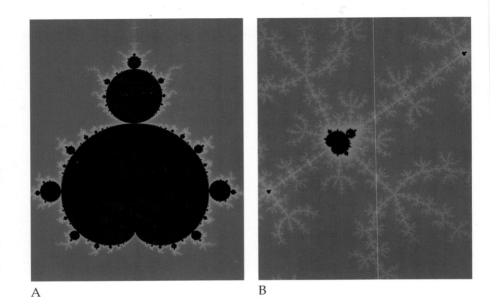

A   B

Mandelbrot set generated by Randall
Kay Stokes on a Hewlett-Packard
9000/370 computer with a monitor
resolution of 1280 x 1024.

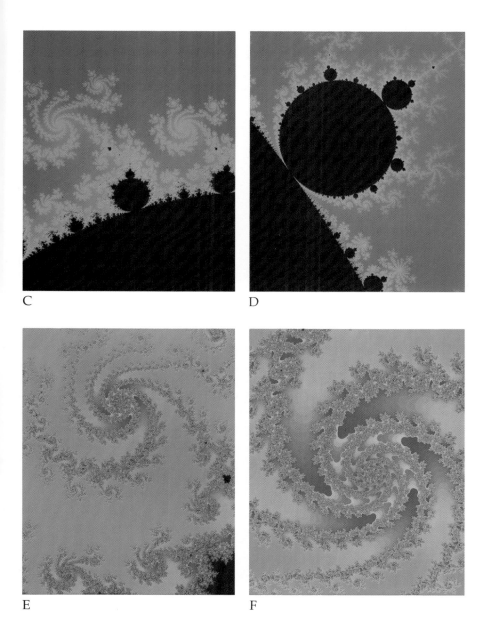

C

D

E

F

2

The Mind Model

For simplicity's sake, I made my topology of the mind in the shape of a cube. I took each part or aspect of the mind and gave it a phase-space within the cube—I changed its "shape," without changing its "function," and thus was born the Mind Model. This model has proved to be the most helpful, the most universal tool for visualizing, understanding, and working with the dynamics of the mind that I have ever seen.

The Mind Model has been tested over the course of many years. I've watched how it has worked in each case, revised it, fine-tuned it, and simplified it—without diminishing its power or effect. The more it's used, the more powerful it becomes. It makes clear how the mind integrates and organizes itself according to the Implicate Order of the universe—and how the mind develops according to this "order within all orders." The Mind Model shows you how to access the Quantum Wave, use the Quantum Force, and unfold the "I" within you. It takes you into the deepest orders of your mind and stays with you to the furthest reaches of the holodynamic universe. It is the greatest gift that I can share with you.

The Mind Model has the following dimensions:

Rational and Intuitive Processes

There are two fundamental ways of thinking, two mental processes that account for almost all we say, do, and feel. One is rational. The other is intuitive. One reflects the "particle" aspect of reality, the other the "wave" aspect of reality.

The rational process takes place mainly within the left hemisphere of the brain. Here can be found our ability to count and calculate, to think in linear, logical sequences, and to analyze—to detect details or "particles." Rational thinking allows us to take the "particles," count them, and find some rational relationship between them. It makes literal sense of all the parts. Within the left hemisphere, we do most of our everyday thinking: we balance the budget, read the newspaper, handle the ordinary tasks of daily life.

The intuitive process of the mind takes place mainly within the right hemisphere of the brain. This hemisphere deals with the "wave" aspect of reality and is home to our imagination, creativity, and artistic abilities. Here our mind dreams, makes love, sees "the whole picture," and feels with emotion. Here can be found our ability to phase-space, to see the world as an interacting, dynamic whole, and to perceive potentials. Here resides the "I," the Full Potential Self. And here are the keys to opening the door to the unlimited potentiality of the Quantum Wave.

These two hemispheres also control our bodies. Split-brain experiments, where Roger Sperry and others cut the corpus callosum (the band of nerve fibers connecting the two hemispheres) in order to save the lives of certain epileptics, have shown that the left hemisphere of the brain controls the right side of the body, while the right hemisphere controls the left side of the body. Thus, writing is done with the right hand (in most people) because it is

RATIONAL MIND	INTUITIVE MIND
"Particle"-focused: picks out the notes in a melody, the letters in a word; takes everything apart.	**"Wave"-focused:** hears the melody, gets the message in speech or writing; puts everything together.
Analytical: analyzes everything, wants to know all the facts and details, to find a logical explanation for everything.	**Holistic:** combines and connects; processes all types of information simultaneously; sees the "whole picture"; makes quantum leaps.
Literal: gives an exact meaning to things; wants a place for everything and everything in its place.	**Metaphoric:** gives deeper meanings, connotations, and images to things; allows for new possibilities.
Verbal: forms and processes words: your talker, reader, and writer; recalls names and dates; spells.	**Emotional:** creates feelings, intensity, and passion.
Logical: thinks in ordered sequences: if A, then B.	**Imaginative:** imagines, dreams, makes up stories; knows how to play; wonders "what if..."; senses new possibilities.
Linear: processes things one step at a time; looks at how one thing follows another, or causes another to happen.	**Spatial:** relates your position to everything around you; helps you find your way around the house, drive a car, walk or run "automatically."
Mathematical: focuses on numbers and symbols; counts; looks for rules by which things relate; sees all things as mathematical.	**Creative:** comes up with new ways of doing things; envisions; solves problems: your lover, artist, poet, inventor, musician.

mostly rational, or a "left-brain," activity. But we say something is "from the heart" when our intuitive mind is involved because the intuitive mind functions mainly in the "right brain," which, when activated, stimulates the left side of the body and thus the heart. The heart is one of the most sensitive organs to respond to the right hemisphere: "heartfelt" means just that—our intuitive thinking is literally "felt" in the heart. So some very down-to-earth things in our daily lives are controlled by the two halves of our brain.

Now, your intuitive process, with its "wave" focus, is well adapted to understand quantum thinking and the holodynamic universe—and to solve even your most difficult problems. Not so your rational process, with its naturally limited "particle" focus. Solving problems from a strictly rational point of view is laborious, lengthy, and often unsuccessful. But when you use your intuitive process to get in touch with the thought streams of your mind, to concentrate on the orders within orders, problem solving becomes dynamic, dramatic, easy, and enduring.

The rational and intuitive processes are represented in the Mind Model as on the following page.

Between the rational and intuitive sections of the model, running diagonally through the cube, lies the "holodynamic plane." The holodynamic plane is a multi-dimensional field where all the activated thought-forms (holodynes) of your mind operate and interact, where the synthesis between your rational and intuitive thinking takes place, where your thoughts and behavior are correlated, and where most of your mental "games" are orchestrated.

All mental activity, both rational and intuitive, begins with the senses. Your senses are the measuring tools of your mind, the means by which you experience your surroundings and perceive reality. As you explore how your senses work, you will understand how the balance between both your rational and your intuitive senses is essential to unfolding your fullest potential.

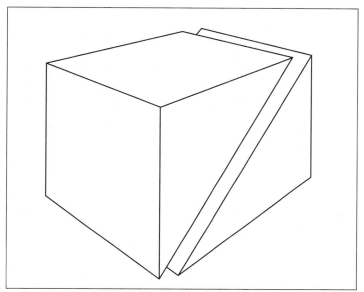

FIGURE 2: RATIONAL AND INTUITIVE PROCESSES

Your Senses and Reality

Your senses give you all your experiences: pain and pleasure, harmony and disharmony, growth and decline. With your senses, you see what you want to see, hear what you want to hear, feel what you want to feel, and taste what you want to taste—light and dark, loud and soft, smooth and rough, bitter and sweet. If you compare your mind to a camera, then your senses are the lenses of the camera, for whatever comes into your mind comes in through your senses.

And when you phase-space your senses you discover all sorts of things that are not yet widely known. For one thing, you learn that all your senses detect both the "particle" and the "wave" aspects of reality. Your *"particle"*-sensitive senses send their information to your *left* brain, while your *"wave"*-sensitive senses send their information to your *right* brain.

Take your sense of sight, for example. In each of your eyes you have a pinpoint concentration of rods on the retina called the fovea. The fovea picks up sight "particles" and discriminates one "particle" from another. The fovea sends its information down the inner fibers of your optic nerve to be processed by your left brain. Around the fovea is the periphery. The periphery takes in the sight "wave"—how you see all the "particles" together, within their visual context. The periphery sends its information down the outer fibers of your optic nerve to be processed by your right brain.

All your senses have this dualistic nature. Take your sense of hearing, for another example. In each of your ears you have a coiled inner organ called the cochlea, filled with fluid and lined with tiny hairs of different lengths, called cilia. Each hair, or cilium, is tuned to a different note or frequency and each picks up a different sound "particle" and sends this information to your left brain to be processed. Not far from the cochlea is the eardrum, which takes in the whole sound "wave," the "beat" of the sound, and sends this information to your right brain to be processed.

Regarding the senses, there is now irrefutable evidence that we all have a "sixth" sense. Evidently, every cell in the body is capable of picking up and transmitting subtle energy in the form of resonating frequencies. At the World Research Foundation's International Convention last year, I met with Bjorn Nordenstrom, one of the world's leading cancer specialists, and heard him present his experiments on the use of the body's subtle energies to heal cancer. Nordenstrom was able to take a cancer cell and, by changing the resonating frequencies of the subtle energies surrounding it, transform that cell into a healthy one. In the last few years, leading researchers on the body have been fascinated by the fact that we all seem to be "walking antennas." We are picking up encoded resonating frequencies all the time. Ernest Rossi reports that the limbic center, at the very base of the brain, is "altered-state" sensitive: it receives, and evidently processes, subtle energy frequencies as they are picked up by the cellular network and sent to the brain. There seems to be a rational explanation for the "sixth" sense after all.

The limbic center is neurologically attached to the hypothalamus, which controls the pituitary, which controls the hormones that trigger your body's responses. When someone says, "Boo!" and you are not expecting it, your limbic system reacts by instantly mobilizing your entire body to be startled. And when you think a certain thought, the holodynes involved set up resonating frequencies, which go directly to the limbic center, which then sends their message on through to the entire body.

Cleve Backster, the inventor of the polygraph, or lie detector, has carried this idea one step further, showing how the sixth sense has a quantum dimension. I first met Cleve at the International Conference on Human Functioning in Kansas in 1987. At that time, he showed a split-screen video recording of one of his experiments. The subject, an obviously skeptical doctor, was asked to rinse his mouth out with a weak salt solution. The solution was then taken into a separate room, placed on a table, and connected by electrodes to a polygraph. The doctor, in the other room, was given a *Playboy* magazine with pictures of Bo Derek in the centerfold. One half of the screen showed the polygraph's readings from the mouth rinse containing the doctor's saliva cells. The other half showed the doctor looking through the magazine. When he reached the pictures of Bo Derek, the polygraph readings went off the chart. Only when he stopped looking did the readings return to normal. The doctor closed the magazine but then, after a moment, said, "Oh, one more look!" and, as he opened the magazine and looked at the pictures, the polygraph readings again went off the chart.

On national television Backster's experiments took an even more dramatic turn. The host of "That's Incredible" showed

Backster taking a single cell from a woman's mouth and hooking it up to an electroencephalograph (EEG), which produces readings similar to the polygraph. With one camera on the readings, the television crew followed the woman with a second camera. The woman walked down a rough section of town at night, was accosted by a pimp, kept on walking, and all the while the EEG readings, even though over three miles away, still recorded her emotional reactions. A single cell can pick up messages in some dimension which is not affected by time or distance—evidently, a quantum dimension.

The implications for space travel are tremendous. It may be possible to maintain instantaneous communication over unlimited distances simply by tuning in to this subtle energy field—by picking up specific coded messages which come through the resonating frequencies of the Quantum Wave.

The evidence, then, is clear that we have six senses to help us. It also appears that each of our six senses has two ways of perceiving reality, either as "particle" or as "wave." "Particle" information is processed primarily in the left hemisphere of the brain and "wave" information primarily in the right.

And how does your mind handle all this information? It takes both the "particle" and the "wave" information coming into it from your six senses and puts it together on the holodynamic plane, where it can see all the information as one, dynamic whole. To understand how the holodynamic plane works, you need to learn about I.S.P., Intuitive Sensory Perception—the way your intuitive mind uses your six senses.

Intuitive Sensory Perception

The key to understanding your holodynamic mind rests in your Intuitive Sensory Perception (I.S.P.). When, through phase-spacing, you tune your senses in to the orders within orders of your mind, you are using Intuitive Sensory Perception. I.S.P. can help you understand the dynamics of these orders within orders: it can help you unlock and unfold your fullest potential, and thus empower your life. This one skill, the ability to use your I.S.P., can do more for you, can do more to get you from point A to point B, than any other single skill.

I.S.P. is responsible for almost all our technological advances and may very well be the key to the future of peace on earth. You are already using I.S.P. Indeed, everyone is, but few know anything at all about it. Here is how I.S.P. works.

Intuitive Sensory Perception refers to the *intuitive* use of your senses. In other words, you take your senses and use them to go "in-tu-it," or into the "wave" function of reality, to experience things in specific, intuitive ways that are not directly available to your rational mind.

Just as physical reality has both a "wave" and a "particle" function, so your senses are able to pick up both the "wave" and the "particle" aspects of reality. You all know, for example, how to use your senses to tell what's cooking for dinner. You go into the kitchen, lift the lid on the pot, look in, take a deep breath, and let everyone know how good it smells, and you might, if the cook will allow, even snitch a quick taste. These are all examples of "particle-focused," or rational, use of the senses. The information picked up is sent directly to the left brain.

What, then, is the *intuitive* use of the senses or I.S.P.? I.S.P. allows you to phase-space into an entirely different dimension of reality—the "wave" dimension. I.S.P. allows you to sense the whole picture, to directly access the thought-forms and energy fields contained within any given circumstance—your own or anyone else's—and to directly intervene in their dynamics. With I.S.P. you can "talk to" the various "shapes" which are controlling the "streams" of thoughts and feelings deep within your mind— and directly relate to them through one or more of your senses.

When we look at a Mandelbrot set, we can see that even in a simple mathematical formula there are orders within orders within orders. We learn from computer re-creations of these orders how beautiful they can be, and also how the "shapes" (the "turtles") interwoven deep within a "stream" can determine the flow of the whole "stream." When we change these shapes we change the whole flow of the "stream." So the "shapes" *control* the flow of the "stream."

Such is the case with many of nature's streams. The rocks within a mountain stream affect the flow of the water and determine its path. The shape of a mountain range affects the flow of air currents—the airstream—and determines the path of an oncoming summer storm. And so it is with the human mind. For the "streams" of thought which govern our behavior, and which dictate to us the way we perceive things, are all influenced and controlled by "shapes" within the orders of the mind. These "shapes" I call "holodynes."

Holodynes

Holodynes are thought-forms which have the power to cause things to happen. Holodynes are single, self-sufficient units of the Quantum Force, fundamental units of the one, dynamic whole—the holodynamic universe. Holodynes respond to all dimensions of the Mind Model. They have both a "particle" and a "wave" function—with direct access to both physical and quantum dimensions, they align themselves within the six stages of development of your mind, and they control the interest wave—and thus your thoughts and feelings.

If you would like to experience a holodyne, just think of your grandma. What color of hair does Grandma have? Can you "see" Grandma in your mind's eye? The reason you can imagine

Grandma (or Mom or Dad) is that the memory is stored as a holodyne. Your memory of Grandma is a living, powerful influence in your life. Often, unbeknown to you, you will love the way Grandma loved, think the way she thought, and sometimes make decisions intuitively, the way she would. She may not be the only holodyne involved in these activities, but there is no denying the impact Grandma (or Mom or Dad) has had on your life. This is one way holodynes program your life.

From a "particle" perspective, holodynes are specific thought-forms that make up our memory banks, the computer programs for our behavior, and the power packs behind our feelings. According to Karl Pribram, the way we receive messages from our senses, send these through our central nervous system, and then store them in our brain works a lot like a hologram. Holograms can tell us much about how we perceive, think, and remember— and about holodynes, which resemble them.

HOLOGRAMS

Holograms are formed by splitting a laser beam so that part of it bounces off the object you want to photograph and part of it bounces off a photographic plate. When you shine a laser through the plate, the holographic image, or hologram, appears in three dimensions. The amazing thing about a hologram is that if you break it into pieces and shine the laser beam through any one piece, *the whole image* appears. From such experiments comes what is called the "holographic principle": what is known to the part is known to the whole and what is known to the whole is known to the part.

MEMORY STORAGE

Memory storage works on a similar principle. Within the neurons of the brain there are no holographic plates, but there are protein strings, and memory is formed and stored within these strings in much the same way as holograms are stored on holographic plates. I call these holographic memory storage units "holodynes." Holodynes behave like living holograms.

CAUSAL POTENCY

Holodynes have power. They can cause things to happen. Roger Sperry won the Nobel Prize in 1978 in part for identifying how thought-forms within the mind develop "causal potency," the power to cause things to happen. According to Sperry, causal potency is created in the mind as a bioelectrical buildup, like a charge in a battery. The more you charge the battery, the more power it develops. This is one rational way to explain how holodynes can develop power.

Another way to explain how holodynes develop power is to look at how your conscious focus affects someone's behavior. If you choose to focus on the negative aspects of a person's behavior, you actually create more power for the holodynes which are causing the negative behavior: the person behaves *more* negatively, not less. Your being negative about someone else's negative behavior simply multiplies the power that the negative holodynes have. It sets up a negative "field." It gives off negative "vibes."

To understand how a holodyne is formed, compare your mind to a camera, and your senses to the lenses of the camera. Imagine your mind reaching out into the world and receiving information back through your senses. This information is sent through your central nervous system, where it is prepared for storage in your brain as an image, a holodyne. The process might look something like the illustration on the following page.

Once in the fluid dynamics of your mind, the holodyne begins to function as a part of your mental activity. Some experiences will increase its causal potency, others will not, or will leave it dormant, depending upon your choice and your focus.

For example, screaming and shouting at a child who is crying

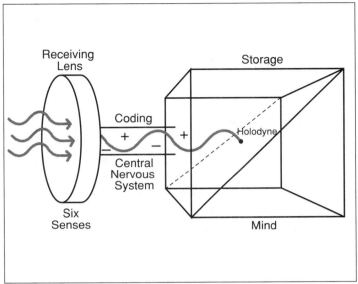

FIGURE 3: THE FORMATION OF HOLODYNES

does not decrease the crying. Nor does hitting a child who has just hit somebody else diminish hitting. Such behavior simply empowers the hitting holodynes, so that love becomes hitting, negotiating becomes hitting, play becomes hitting. Hitting becomes integrated within the mind with all kinds of activities and eventually, collectively, we can only "get along" with our neighbors by striking out at them.

Such is the case of the Iks of Kenya. A nomadic tribe of hunters, the Iks live in the middle of Kenya National Park. When the government established the park, the Iks were forbidden to hunt there anymore: all hunting became poaching. The Iks were forced by governmental decree to become an agricultural society. Their response was to withdraw completely into themselves. They have become a schizophrenic people. They only laugh when one of their neighbors is in trouble or in pain. They ignore their young and abandon their old. They defecate on each other's doorsteps and hurl insults at each other all the time. This is now their culture. It came about because their individual holodynes of anger, frustration, and hatred could not be collectively expressed. So the Iks' society broke apart into individualized expressions of anger.

When you phase-space such expressions of anger and then look at our society to see how *we* express anger, some interesting dynamics turn up. You can see, for example, that collective groups often do exactly what the Iks do individually. Does not one high school team laugh when another fails? Does not one city defecate on another's doorstep by allowing its pollution to pour unchecked into the air and water of another? Do not governments—local, state, and federal—still tend to abandon both young and old? Our collective mentality is still evolving and we must learn to phase-space it, so that we may help it develop in a systematic way and unfold our collective potential.

The principles and practices of fairness, peacefulness, and cooperation can only be established if the holodynamic arena is in harmony. The primary units of the holodynamic arena are the holodynes within each of us. Society is like an orchestra where all the individual players must learn the score before the symphony can be played. Using I.S.P., we can each "learn the score" regarding our individual holodynes and can take responsibility for the causal potency of each, so that our individual state of being adds to the collective harmony of the group. The peace we create within ourselves, among the holodynes of our minds, is the real beginning of world peace.

The Enfolded "I"

It is within the phase-space of holodynes that the "I" resides. Once you learn how to phase-space, you can learn how to access the inner world of your holodynes. And once you learn to phase-space what you really want, once you're no longer "buried in the ball of string," you can see how focusing your mind on the Quantum Wave, the whole range of all potentiality, helps you unfold any specific potential, and get what you really want. You can see that the key to getting what you really want is enfolded in your Full Potential Self. This is the great secret. This is what most successful people have learned. Your Full Potential Self, your "I," knows how to tap into the Quantum Force, access the source of all things possible—the Quantum Wave, and unfold your enfolded potential.

But unfolding your Full Potential Self involves clearing the channels of your mind of all accumulated debris—all your immature and fractured holodynes. It requires changing the "shapes" within your mind, changing the very nature of the holodynes that are controlling and limiting the "streams" of your thoughts, feelings, and behavior. The holodynes that limit your thinking and curtail your potential are immature. When you access such holodynes and mature them, you change their order, and likewise, all other orders. A Mandelbrot Set, in all its complexity, is governed by a simple formula, which has a corresponding, governing shape. What I learned early in my experiences studying the human mind is that the mind also has its own "formulas" and corresponding "shapes." And these "shapes," or holodynes, control the order of our thoughts.

If you were to focus for just a moment on why you're doing what you're doing, and, more importantly, on what you really want to do—what you really want in life, you would see that the potential hidden within you keeps reflecting outward. You would see that all of your wants are simply reflections of your hidden potential, which keeps saying, "I want out. I want out."

The things that you seem to want "out there" in society are really potentialities within you. Now, your mind has several ways of handling this. One of the ways is to get angry and to start "clubbing" everybody around you. Another is to shift and try to change your external circumstances. This is a searching, or "trying-to-find," process. But if you just go inward, you will find that what you really want is hidden in an order within you.

When you feel turbulence and chaos in your mind, when you are filled with turmoil and you feel like you're going to "lose it," you can phase-space your turbulence and you will find that the turbulence is there, as everywhere, because new orders are evolving. And if you look beneath your chaos to find the holodynes involved, you will soon see the lessons it has to teach you—the intent of your chaos—and you can then open the doors to the new order which lies within it. You can phase-space your whole mind and see that it is a near perfect reflection of the holodynamic order of the universe.

Are you looking for a better, more intimate relationship with someone? Do you have the potential to love more deeply? Do you have to have a ready-made model "out there," one just "made for you," so you can love and have someone accept your love? Or is love the process of unfolding the potential for intimacy within you and investing that in a living relationship? How can you tell, how can you find these orders within orders? What is controlling the orders? And where does your potential to love come from? How can you change what has gone on for generations in your family and culture?

The answers to these questions can be found in the world of holodynes, in the hidden orders of your mind which control the things you feel and do in your everyday life. In a way, phase-spacing holodynes is like scuba diving. Underwater, a whole new world opens up to you—a world you never knew was there, just beneath the surface, a world of wonder, beauty, and adventure. So let's "dive in" and look at the world beneath the surface of your conscious mind—the world of holodynes.

Family Holodynes

The most powerful, permanent, and potentially devastating holodynes that you ever get come from your family. From the moment you are born, your family imprints you with its beliefs, whether you like them or not, and these beliefs form the images that govern how you think and feel, and what you do. Some of these beliefs are passed along to you genetically, as split-twin studies the world over have shown. Indeed, it seems that some imprinted family beliefs are passed on as holodynes again and again, from one generation to the next.

But the most obvious way that you get your family holodynes is by modeling. You see abusive love in action, and you become an abusive lover. You hear family members say that they are "dumb" or "can't succeed," and you grow up believing that you, too, are "dumb" or "can't succeed." Your potential becomes *blocked* by this limited thinking, *blocked* by the holodynes—especially your family holodynes—that control the dynamics of your everyday life. And your potential percolates within you, trapped like boiling water inside a pressure cooker. This is why so many people have nervous breakdowns, why so many have rashes and diseases of all kinds.

I have assigned family holodynes, which are active and interactive in one part of the mind, to the specific area of the Mind Model entitled "Family Beliefs," on the following page.

Into this area go all of the beliefs you got from your family. Can you think of things your family taught you? What, for example, did they teach you about food? About manners? About language? What did you learn about good and bad? About how to deal with people? The meaning of life? And what did your family teach you about love? Justice? Health? What did they believe about work? Money? Power? What were your family's attitudes toward sex? Secrets? And success?

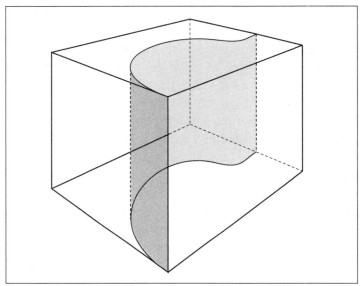

FIGURE 4: FAMILY BELIEFS

Whatever your family believed about all of these things is stored within your mind in the form of holodynes. When you first began your life, you were like a computer without any software. You had few developed holodynes, so your active mind had no way to manifest itself. But as you went along, you gained more and more holodynes and, especially at first, during the first few years of your life when you were most impressionable, these were provided by your family. Your mind became partly filled with the belief systems of your family. You live your life by them.

If you are like most people, towards the middle of life you begin to re-evaluate your family belief systems and ask, "Do I want to keep these or not?" You go through what is sometimes called a "mid-life crisis" or a "mid-life passage." To be your own person, you must deal with your family holodynes, both individually and collectively and which, more than likely, are the most powerful holodynes in your life.

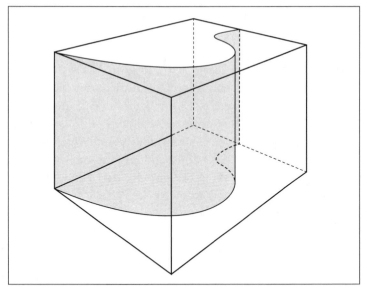

FIGURE 5: CULTURAL BELIEFS

Cultural Holodynes

Family belief holodynes are not the only belief holodynes that affect our lives and control our thinking and behavior. I have taken all the other belief holodynes in society, called them "cultural beliefs," and placed them on the other side of the Mind Model, across from family beliefs.

Contained in the area marked "Cultural Beliefs" are the holodynes you have gotten from your peers or associates, from your social or cultural groups, and from society. Here are all the things you learned in school, at work, in church or temple, in clubs or associations, and in all your interactions with society.

Most of these holodynes are very carefully passed on from generation to generation by society itself. The educational institutions, for instance, pass on certain widely-held values and beliefs. Can you remember your first day at school? Think back for a moment on what you learned. If you were to pick a half a dozen things you learned from school, what would you pick? Some people learn to stand in line. Some people learn to raise their hand before they talk. Others learn when to go to the bathroom and where. Some people learn to color, to write, to count. Some people learn to fear authority figures. Others learn to love their teachers. What holodynes come to your mind when you think of school?

Then think for a moment about your peer groups. What did you learn from your friends? When did you receive your first real kiss? How did you feel? What did you decide?

What have you learned from work? What lessons have you learned from your career choice? What have you learned from your government? What do you believe about politics? The holodynes governing all of these beliefs are contained within the cultural beliefs section of the Mind Model.

Comfort Zone Thinking

Now, note how wonderfully correlated your mind is. Your mind takes all your family beliefs and all your cultural beliefs and it overlaps them. This "overlap" between family and cultural beliefs is called your "comfort zone," as seen in the Mind Model on the following page. Into this area go the holodynes of beliefs your family and culture agree upon, beliefs you find "comfortable," stable, and consistent enough to accept as "true."

Take, for instance, the Nyars in India. The Nyar men run around playing jungle war games with bows and arrows. They don't hurt very many people, but they have a lot of fun and that is their lifestyle. Nyar women, on the other hand, live back in the villages, they build the homes, raise the animals, rear the children, and take care of everything having to do with domestic life. It's a matriarchal society. Every woman has seven or eight husbands, and at least as many lovers. The men only come into town once in

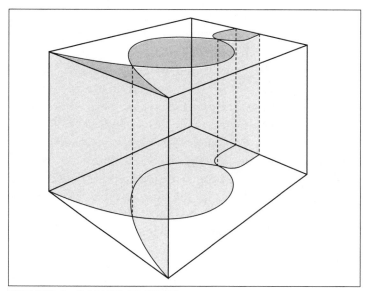

FIGURE 6: COMFORT ZONE

a while, and when they do, they line up in front of their spouse's or lover's door and present her with gifts, and it's all controlled by ritual and tradition. And that's how Nyar children are conceived. The men go back to their jungle games and nobody really knows who fathers the children. They have a word for all this and it means "God's chosen way." Because they believe firmly that all people ought to live this way.

Every society has within it organizations that believe all people ought to live *their* way because that's "the way God commands it" or that's "the way God would want it." Almost universally, human beings tend to take their comfort zone thinking and "divinize" it, with the usual result that they no longer have to be responsible for it. It's "God's chosen way." Divinizing war means you can no longer challenge it. It's non-negotiable. It becomes the greatest way there is to die. Thus a whole community of believers becomes comfortable with a warmongering theology. And thus are planted the seeds for the "holy wars" that we can no longer afford.

The same is true in the economic and political arenas. As our technology grows more advanced, and our weapons "improve," we can less and less afford the negative luxury of "divinized" comfort zone thinking. The time has come to phase-space not only our cultural belief systems, but all of the religious, philosophical, economic, and political beliefs within them. We need to look at the real impact such beliefs have upon society and upon the world, and to *mature* the primitive comfort zone thinking that has created wars in the past—and that is still creating them. Our future depends upon it. We can begin the process by phase-spacing these beliefs within the framework of the Mind Model.

Even though comfort zone holodynes are among our most powerful, motivating holodynes, even though our comfort zone beliefs simply *feel* true, they are *not* true—not in the way we perceive them to be. Phase-spacing can universally help us realize the deeper meaning of the metaphors we use to perpetuate man's inhumanity to man. But it means getting *into* and *out of* our comfort zone thinking.

Once we understand that family and cultural belief systems are actually passed on by holodynes, we can begin to look at how the mind stores memories as holodynes. And we can begin to see, in those holodynes, the orders within orders of the mind. Every holodyne aligns itself within the resonating frequency of its own stage of development.

The Six Stages of Development

The six stages of development of the mind reflect the Implicate Order of the universe within everything that grows. Thoughts, feelings, and actions—the mind, human beings, societies, and the universe itself—all develop according to these six stages. Although my formulation of them is unique, these stages have been observed and described by some of the world's great researchers on human development, such as Jean Piaget, Lawrence Kohlberg, and Abraham Maslow. And they are clearly implied in the writings of philosophers and religious thinkers throughout history. The six stages of development of the mind are as follows:

STAGE ONE: PHYSICAL

In the first stage, the mind becomes aware of its physical surroundings and starts to make sense of things. Here we learn to perceive and to experience the physical world—what it means to be alive. Here we *become* physical.

STAGE TWO: PERSONAL

In the second stage, the mind becomes conscious of its own identity, its personal self. Here we learn to be persons. We become self-aware and self-confident. We learn to assert ourselves and to play all the games that people play about self-esteem. We learn about personal creativity.

STAGE THREE: INTERPERSONAL

In the third stage, the mind recognizes other people as other "selves" and enters into living relationships with them. Here we learn to relate, to develop rapport, mutual respect, and friendship. We learn about intimacy: we discover how it feels to commit to another person, to become part of that person's life.

STAGE FOUR: SYSTEMS

In the fourth stage, the mind shifts into "systems thinking": it can see how human beings work together in all kinds of systems. The "company mind," the "church mind," the "family mind" and the "team mind" are examples of this thinking. Here we learn to cooperate with groups of people *beyond* our intimate relationships; we learn about open trust and teamwork. And, as group members, we learn to live our daily lives, to act out our part in society.

In the fifth stage, the mind comes to understand, and to identify with, the principles by which systems live and grow. Here we learn about love and faith, compassion and integrity. We learn to live these principles, to own the circumstances we have created for ourselves, and to "become" love, faith, compassion, and integrity.

In the sixth stage, the mind that understands the living principles of systems comes naturally to universalize these principles, to apply them to *all* people everywhere. Here we learn to live as one with the universe. Not only do we then love our neighbor, but we consider *everyone* our neighbor. We become attuned to all of life. We become the manifestation of love itself as we interact with the world. We experience the fully unfolded "I" and extend ourselves harmoniously into the holodynamic universe.

It is toward this universal mentality that humanity is evolving and has been evolving over history. All holodynes, all minds, all human associations and systems evolve according to the stages of development: from individuals, personal relationships, families, and companies to countries, cultures, principles, and humanity as a whole.

At each stage in the mind's development from immaturity to maturity, holodynes form and align themselves with other holodynes which have similar characteristics and use similar processes. They take on responsibilities and follow general principles. Through their "fields of influence," which extend outside the

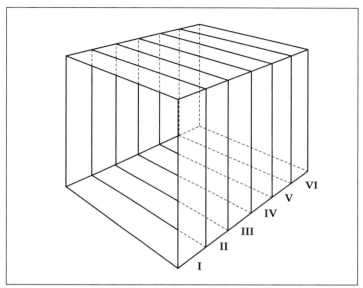

FIGURE 7: SIX STAGES OF DEVELOPMENT

mind, holodynes create your patterns of behavior and shape your life circumstances. I have represented these six stages in the Mind Model as shown above.

The Six Stages of Love

Take, for example, love. When you phase-space love, using the Mind Model as your frame, you can clearly see love's six stages of growth:

Love can be physical, and strictly physical. At Stage One, your love is purely an expression of physical pleasure and vitality.

Love can also be the love of self. At Stage Two, you love yourself but you have not yet learned to love others: you value only what is best for *you*.

Love can be the love of an intimate relationship. At Stage

Three, you love the other person and the living relationship as much as you love yourself. You are there for the other and allow the other to be there for you. Your love means commitment, loyalty, mutuality, bonded oneness.

Love can be the love of a whole system. At Stage Four, you love your country, congregation or school, your social group or business organization. This stage of love unfolds in an atmosphere of cooperation, open trust, and teamwork, when people really care for one another, as a group, and experience the sense of triumph that comes from accomplishing difficult tasks together.

Love can be the love of principles: Stage Five. When, for example, you get "into" love as a principle, you realize that "I, myself, *am* love," that "love manifests itself through me." A deep sense of integrity unfolds within you. You "become" love.

And, finally, love can be universal love. At Stage Six, you understand that everyone everywhere is manifesting universal love. You become at one with all humanity, all life. You extend your love throughout the universe.

Now, when you come into life *deprived* of love, you grow up feeling deprived, you come to think of yourself as someone who always "needs" love, who can never "get" enough love. You will usually seek out someone who is "nurturing" to "give" you the love you "need," but who, in fact, "needs" love as much as you do. And you will play the game of "needy" and "nurturer" until one or both of you get tired of it. Being deprived, then, at Stage One becomes defining yourself as "deprived" at Stage Two, which becomes entering a "deprived" relationship at Stage Three.

Sometimes, at Stage Four, you will naturally join forces with a whole group of "nurturers," who, of course, also "need" love. Entire organizations can be based on the "need" to "nurture." It is not until you get to Stage Five and you realize "I *am* love," that you can begin to phase-space the "need" for love and understand more clearly the dynamics involved in all the games people play about love.

Picture, if you will, a little leech swimming along in society, saying, "I need love, I need love," and then picture another one coming the other way, saying, "I need love, I need love." When the two leeches meet, they cling right onto each other, but you can see that this kind of relationship will not last long: there aren't any nutrients in it, just two little leeches who "leech off" each other.

But when, at Stage Five, you understand that you *are* love itself, and when you can own that, you "become" love, and you never "need" love again. You understand that you can give love, resonate with love. You can "cast your bread upon the waters," and it will come back to you. You can plant the seed of love and it will grow up ten times stronger again.

So, as you phase-space love, you can see the six stages of love and how love evolves from stage to stage. You can see how physical love becomes self-love, which becomes love of another, which becomes love of systems, then love of principles, and finally universal love. Using the Mind Model, you can identify the patterns of love you practice in your life —with your friends, with your family, and with yourself: you then have a greater degree of freedom in choosing which pattern of love you would like to focus upon, and what to do about blocks to the fullest expression of your love.

The Six Stages of Trust

You can explore any subject in terms of these six stages. Take trust up through the six stages, for example, and begin to explore the patterns of trust you have developed at each stage. How do you trust in physical things? In yourself? How do you trust in relationships? How do you trust in systems like the government or where you work?

Sometimes you develop patterns of trust in the early stages of your development which do not allow you to unfold your fullest, most mature trust in the later stages. You cannot understand, for instance, the fundamental principle of trust—that you *are* trust. You learn the lessons of blind trust, but not those of *real* trust, open trust, which says, "Let's put everything on the table." Or if you can "understand" trust, you choose not to *own* it, make it part of all your dealings. You somehow mistrust trust, and so, of course, it doesn't "work" for you. And when you still get "taken," you mistrust trust all the more: you hide from others—and from yourself. Only when you choose to "become" trust, to own it, to make it part of your very state of being, will it work for you.

What, then, is universal trust? It is a living dynamic with great potency, great power, which naturally inspires people to work together for a common good. Most truly successful organizations in the world draw upon a deep, abiding, dynamic reservoir of universal trust. Their people, from chief executives to filing clerks, live by ethical principles they would not think of violating. They can count on one another. They work as a team.

How do you develop the trust which inspires teamwork? How do you prepare your mind so that you and your teammates can work synergistically to your maximum potential? You start by phase-spacing. You phase-space your mind, and the minds of your teammates. You phase-space the situation you're in. You look at the level of maturity and the patterns of thought involved and then you focus on the potentiality. And in doing so, you take the whole dynamic of trust through the six stages.

What you find is that deep within the streams of thought of your mind are hidden orders within orders. You find that, within these orders, all your holodynes align themselves to form ordered patterns of thought and personality traits. And you find that you can influence and control these patterns by your conscious focus and choice.

Upddraft and Downdraft Choices

In the chart entitled "The Six Stages of Development," I have outlined a general, phase-spaced review of typical thought patterns and the crucial choices which occur at each of the stages of development. Note that, depending on a crucial choice at each stage, the mind flows either *upward*, towards order and well-being ("updraft"), or *downward*, towards disorder and dysfunction ("downdraft"). At each stage, distinct mental patterns appear, which act as subtle attractors or controllers within your streams of thought.

Your mind is sensitized to a deep order of organization—the Implicate Order. Each experience, stored as a holodyne, gives off a resonating frequency which aligns that holodyne with others of similar or harmonious resonating frequencies. Your mind is like a complex filing system. For example, all holodynes concerned with your physical well-being are "filed" or aligned within the first stage of your holodynamic plane; all holodynes concerned with your personal well-being—your personal growth, your emotions, and your self-esteem—are "filed" or aligned within the second stage. This alignment occurs at each of the six stages and is briefly outlined in the chart.

Some mental patterns tend to unfold more of your potential, others tend to constrict it. The tendency to progress or retrogress—depending on your updraft or downdraft choice—was unique at each stage of development. This proved to be particularly clear in therapy. Whether someone solved a problem or remained hopelessly cycling in its patterns depended upon certain subtle but crucial choices.

These crucial choices determine—at each stage of your development—whether you shift into updraft, and fulfill your potential, or into downdraft, and expend your energies in endless cycles. A downdraft choice automatically shifts your mind into an

CHART 1: THE SIX STAGES OF DEVELOPMENT

I	II	III	IV	V	VI
PHYSICAL WELL-BEING	PERSONAL WELL-BEING	INTERPERSONAL WELL-BEING	SOCIAL WELL-BEING	PRINCIPLED WELL-BEING	UNIVERSAL WELL-BEING
Vitality Abundance Health Strength Energy	Creativity Confidence Self-assertion "I am OK" Self-discovery	Intimacy Friendship "We are OK" Mutual respect Rapport	Synergy Teamwork Open trust Comradery Cooperation	Integrity "I am" Owning it Fair-care-share Openness	Oneness Knowing Empowered Loving Attuned
CHOOSE To live / Not to live	CHOOSE To unfold / Not to unfold	CHOOSE To commit / Not to commit	CHOOSE To act / Not to act	CHOOSE To become / Not to become	CHOOSE To extend / Not to extend
Deprivation "Dingbat" role Dis-ease Zero power Shut down	Denial Fear Anger Insecurity Self-defeating	Disconnected Manipulator "Match-my-images" Pleaser Victim	Conformist "Shoulds" Rule-bound Role-bound Judger	Rationalizer Pretentious Hypocrite Unethical Unscrupulous	Detached Remote Aggrandized Obsessed Tyrannical
PHYSICAL DISORDER	PERSONAL DISORDER	INTERPERSONAL DISORDER	SOCIAL DISORDER	PRINCIPLED DISORDER	UNIVERSAL DISORDER

immature holodyne, activating a negative mode of thinking. This brings about the negative consequences that naturally follow such thinking. With an updraft choice, exactly the opposite occurs. The holodynes you choose, whether mature or immature, determine whether you will handle things maturely or immaturely, whether you will attract positive or negative experiences.

No matter which you choose, updraft or downdraft, you can learn from the consequences of your choice. The lessons of life are laid out for you by the holodynes you choose. It is all part of the Implicate Order. You will continue to experience downdraft until you learn your lesson and choose to shift into updraft.

To choose a downdraft holodyne at any stage is to lock into thinking that blocks growth and begins a cycle back to more primitive thinking. To choose an updraft holodyne is to choose growth, to progress toward personal fulfillment and the unfolding of your Full Potential Self.

The choice between an updraft or a downdraft holodyne may be very subtle, but the results are powerful. Once you have chosen a downdraft holodyne, you may find it difficult to shift into an updraft one. Once you become locked into an empowered, immature holodyne, you must use your I.S.P. to access that holodyne, learn the specific choice or choices which invited its downdraft thinking, and make a conscious effort to mature it into an updraft holodyne. No permanent change will occur in a primitive holodyne—it will maintain its negative causal potency—until it is allowed to grow through the natural stages of its own development. It makes no difference if you think of other things, sublimate, affirm, or meditate. A primitive holodyne retains its power until it is matured.

The crucial updraft-downdraft choices differ from one stage to another, according to the Implicate Order of our creative intelligence. And remember, just as each of our thoughts reflects both the "wave" and the "particle" aspects of reality, so each of our choices has both a rational ("particle") and an intuitive ("wave") aspect. Listed below are the crucial choices at each stage of development and some of the dynamics which surround these choices.

STAGE ONE:

At the physical level, the choice is *to live or not to live*. When you choose to live, you invest your life energy in a specific potential, and thus manifest this potential in "particle" reality. When you choose *not* to live, you remove your life energy back into "wave" reality so that physical manifestation is no longer possible. This occurs whenever you choose not to support a given thought, person, or cause. When, for example, you decide to withdraw from a relationship, you withdraw your life energy from that relationship and the relationship can no longer exist.

Choosing *not* to live is always downdraft because you are no longer "response-able." You cop out. This, or course, does not mean you must never say no to anyone or anything. "No" can be an updraft decision. Turning down something can be a mature choice to turn *toward* something else. When, for example, you turn down abuse, you turn toward a more mature way of relating. When you choose to overcome your addictions and to take charge of your life, that is definitely an updraft choice.

Each time you face some challenge, each time you come to a fork in the road, there is a part of your mind which asks, "Do I really want this?" At a deep level, your mind considers the alternatives and then you choose—to updraft or downdraft. The ability to choose is what makes us intelligent beings. Your body, your personality, your relationships, the organizations to which you belong, and the principles by which you live continue to exist because you have decided to give them your life energy. All of us have been choosing the kind of world we want for so long that the process has largely become subconscious. From a quantum perspective, we have been creating our physical universe by our choices about physical well-being.

Plate II

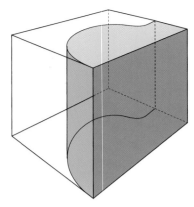

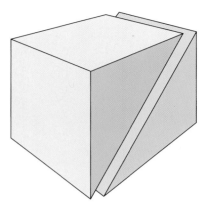

Rational and Intuitive Processes

Family Beliefs

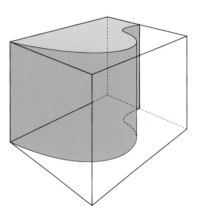

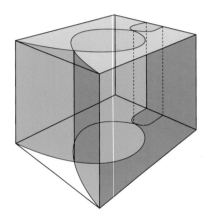

Cultural Beliefs

Comfort Zone

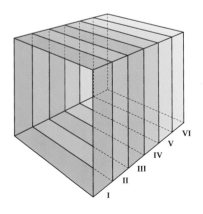

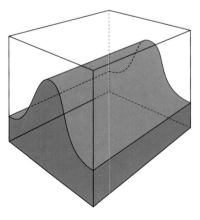

VI
V
IV
III
II
I

Six Stages of Development

Interest Wave

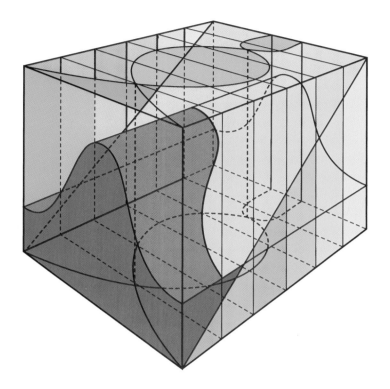

Complete Mind Model

Illustrations by Kathleen Koopman

STAGE TWO:

Once you have decided to put your life energy into something, you must then decide whether or not you will *unfold* your own potential with regard to that something. You must choose whether or not to put your *personal* energy into it. At this stage, the crucial choice is *to unfold or not to unfold*. The personal caring or love you put into any life effort is crucial to the outcome of the effort—and to your personal development. Your love is essential in unfolding the "I" within you, the part of your mind in charge of your personal well-being. Will you personally care enough to solve the problems which arise in the course of your effort, or will you simply sit back and let whatever happens happen?

STAGE THREE:

Out of the first two choices, a third naturally evolves. It is the choice *to commit or not to commit* to a person or project you care about and have decided to put your life energy into. Commitment means taking personal charge of your life energy and *keeping* it *focused* on the relationship or the project. In a relationship, commitment means loyalty and dedication to the living bond between both persons. On the job, commitment means staying true to the agreed upon assignment until it reaches its fulfillment. Enduring commitment promotes the flow of your personal life energy, which, in a loving environment, allows the fullest potential of each person and of the relationship to be realized.

STAGE FOUR:

The first three stages set the foundation for the fourth and next crucial choice: *to act or not to act*. Because the flow of your personal life energy is almost always influenced by immature holodynes, you must learn to overcome these barriers to your empowered potential as an individual, by acting within the systems of society. Once you "birth" a project, once you decide to nurture it and commit yourself to seeing its potential fulfilled, the consequences are set. You must learn to handle these consequences—to *act* upon them. If you decide to "get rich," you must learn everything it takes to make money and to handle it. If you decide to "get healthy," you must learn everything you need to know to achieve health. And you must teach all your holodynes how to act in harmony with your new objectives.

STAGE FIVE:

The Stage Five choice is *to become or not become,* to integrate or not to integrate what you have learned into your total life. How many of us *rationally* know how to get rich or healthy but have never *intuitively* integrated the knowledge? Any new endeavor requires an incubation period in which all aspects of your inner and outer world come together as a part of your new growth. At that point you realize, "I *am* in charge of becoming rich," or "I *am* in charge of my health." As you integrate these principles within your mind, you come to own them, to realize, "I *am* abundance" or "I *am* health." Your conscious self and your individual "I" come into alignment with your personal place in the Implicate Order of the universe.

STAGE SIX:

At Stage Six, you choose *to extend or not to extend:* to use your new holodynes to their fullest capacity or not to, and thus to curtail your growth. To extend your life energy, to bring any project or relationship to its fullest bloom, you must become fully attuned with the Implicate Order. You have chosen to give life to your specific potential, to personally care, to commit your time and resources to it, to do whatever it takes to complete it, to integrate this knowledge into your sense of being and to "become" it, according to the principles of universality. When you then choose to *extend* yourself, to pour into it your fullest potential—the "I" within you, the universe responds. Your "I" unfolds and your fullest potential is realized.

There is great depth to the human mind. There is also great order. Once you see how the Implicate Order aligns all the thought streams of the mind so that they can systematically unfold through all six stages of development, life becomes easier. And fulfilling your own potential becomes much easier.

For instance, one day I was playing with my kids in the living room and, all of a sudden, there was a heavy pounding on the door. My kids all looked at me kind of frightened. I said, "I'll get it," swung the door open, and said, "Hi." And there stood one of my neighbors, who had just moved into a house three doors down the block. With both hands raised above his head, his face beet red, he screamed at me at the top of his lungs, "Your son is ruining my daughter!"

Now you see, I could have dealt with him at any of several stages. I could have been truly physical, and just punched him out—let him have it before he had a chance to lower his arms and defend himself. Or, I could have said, at Stage Two, "You can't talk to me that way!" I might also have said, at Stage Three, "I know. My sons ruin *all* the daughters in the neighborhood." I could have gone to Stage Four and asked, "Do you know what the Bible says about people who act like you?"

I also could have entered Stage Five, which I did, and owned that this man's intentionality was genuine concern for his daughter. So I stepped up to him and said, "If you're really concerned about the welfare of your daughter, you'll look at how you, her father, are treating her. You are doing far more damage psychologically to her right now than my son ever could."

Now his daughter was thirteen and she was standing about fifteen feet away on the sidewalk, crying. She had on her summer shorts and one leg of her shorts was pulled up, exposing a big birthmark, about six inches across. My fifteen-year-old son was nearby on his bike. As I looked at him, he raised one eyebrow and gave me his look that said, "These guys are nuts, Dad." We looked at the daughter, and she immediately stopped crying.

Come to find out, because she was new in the neighborhood, she had wanted to show two younger girls her dreaded birthmark as a sign of confidence in getting to know them. And as she was showing the girls her birthmark, my fifteen-year-old son drove by on his bike, looked down at the birthmark, looked at her, went "hmmm," and rode right by her. She was so embarrassed that a handsome, older boy should have seen her birthmark, she thought her whole social life had just gone down the drain. She went into hysterics and ran to her father. He did not know what the problem was, but that it had something to do with my son, so he took her by the hand, and he marched right over and pounded on my door.

Now, my nineteen-year-old daughter had had a birthmark removed from her leg. She was gorgeous and had sung and danced on stage over half of the world. She came out and said, "Birthmark! Birthmark?" She took the girl under her arm, walked down the block with her, and explained how she once had a birthmark and what she had done about it. They became good friends, and good neighbors.

So the stage of development at which you approach any kind of interaction will unfold for you the particular lessons it has to teach you. For every stage of development has value, and every choice has its lessons. You must learn the lessons before you can move on to the next stage. The wonderful thing is that *you* get to choose when and how to learn these lessons.

The Interest Wave

Einstein once said that if you could truly focus your interest for three minutes, you would be a genius, able to solve any problem. You can see that this is so if you think of interest in quantum terms. Interest represents that ebb and flow of concern, focus, attention, passion, or curiosity which is the sum total of all the holodynes on the holodynamic plane of your mind. Interest comes in waves, accumulations of all the holodynes that are active at any given moment in your mind. I have represented interest in the Mind Model as a wave passing through the center of the mind cube.

The interest wave is one way to show how the unmanifest Quantum Wave makes itself felt in the physical world. It is controlled in size, shape, and frequency by the holodynes on the holodynamic plane. The more mature the holodynes, the greater their capacity to influence the Quantum Wave and to tap into the Quantum Force.

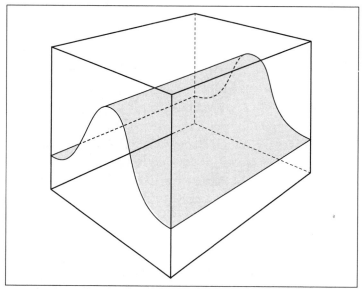

FIGURE 8: INTEREST WAVE

In the realm of quantum thinking, the unmanifest Quantum Wave contains all specific potentials ("quanta") within it. It contains all things possible, but as yet unrealized. This is why some physicists call it the "Offer Wave." It is, in reality, a wave of unlimited potential. Now the Quantum, or Offer, Wave can be compared to a great sea of possibilities, a sea of pure, clear water, and the physical reality of any given circumstance, to a cup of water from the sea. When you focus on this sea of possibilities, when you dip in your cup and lift it to your lips, you collapse the entire sea into your cup: you quench your thirst with that one particular cup of water and with *no other* that "could have been." The possibility—the potential—you choose becomes, *by your choosing it*, a unique and unrepeatable reality.

The Role of the "I"

Your "I" has the ability to go back and forth between the manifest and unmanifest planes. Whenever you focus on—choose—a specific potential, your "I" sees if your choice is aligned with your fullest potential and, if so, carries your focus through to the Quantum Wave. The Quantum Wave responds to your "I" by setting up what is called the "echo wave," which then goes into the past, present, and future, checks out the boundary conditions, and determines if your specific potential can actually become real. The echo wave seems to be the intuitive "wave" function of *every* "I" in *every* dimension—past, present, and future. It's as though all the "I's" know one another and, acting together, either agree or don't agree to allow your potential to become real. If all the boundary conditions are met, the answer comes back: "Okay, what you want can now become reality."

It is because the "I" approves *all* our experiences that there are no victims. Every experience you have ever had has been approved by your "I," your Full Potential Self, and by every other " I."

Putting it all Together

As information is received and relayed by your senses, your central nervous system organizes and encodes it for transmission to your brain. Senses which are *"particle"*-focused send information to the left hemisphere of your brain for rational processing; senses which are *"wave"*-focused send information to the right hemisphere for intuitive processing. This sensory input is then stored within the protein strings of your brain as living, multidimensional images called holodynes, which interact on your holodynamic plane. Here your mind accepts and aligns these holodynes and "files" them for future use, according to a deep, underlying order of life—the Implicate Order. This deepest order arranges all holodynes into the six natural stages of development for the human mind.

All your family and cultural beliefs are stored within your memory banks as holodynes. Your mature holodynes naturally align themselves with each other and become available for use when you choose mature action. Your immature holodynes likewise align themselves, so that you have a wide variety of alternative responses, both mature and immature, to any given circumstance.

Now, when you choose and focus on any given potential, your focus triggers the holodynes aligned with the resonating frequencies of that potential. The Quantum Wave then responds to your focus, once all boundary conditions are met, by releasing the appropriate amount of the Quantum Force, and the potential you have chosen becomes real. In short, the Quantum Wave responds to your focus and choice by manifesting your reality.

When you put all of this together into one model—holodynes, intuitive and rational thinking, family and cultural beliefs, the holodynamic plane, the interest wave, the Quantum Wave, and the stages of development—you have a powerful new tool to use in unfolding your potential: the Mind Model *(see Figure 9 and also Plate II)*.

As you can see from the model, the more mature the holodynes, the more of the Quantum Wave they can manage. All minds respond to the Implicate Order of growth. Not all individuals, however, choose to use their mature holodynes. It is our power of choice that makes us free, or if used unwisely, destroys us.

Some of my "hardest" cases—people that society had completely given up on, murderers, paranoiacs, schizophrenics, and the like, taught me my clearest, most potent lessons. They taught me that nothing is impossible. They taught me that with phase-spacing, the Mind Model, I.S.P., and the use of your mature holodynes, you can bravely go where others fear to tread. That there is a way for you to get from point A to point B, no matter how limited your skills at present. That you cannot even wish for a thing unless the potential is *already* there, enfolded within your "I," your Full Potential Self. And they taught me that, by learning how to phase-space, to "quantum think," to access the holodynes which are causing all the problems in the world, you can confront those problems on the front lines. Everyone has I.S.P. Everyone and everything responds to the Implicate Order.

How, then, do you get to the holodynes that are causing all the problems? You use your I.S.P. to look beneath all the chaotic behavior and you give these holodynes a physical manifestation. Once you "sense" the holodynes, you communicate directly with them, you find their intent, and you mature them to their real intentionality—their fullest potential. This process is called "tracking" and will be explained in more detail in Chapter 3. The tracking process works for all holodynes and can be used to access any problematic or primitive thinking, get to the holodynes involved, and change the physics of the mind. As you reach outward into society, your "I" unfolds and your full potential begins to manifest itself. This process is called "potentializing"

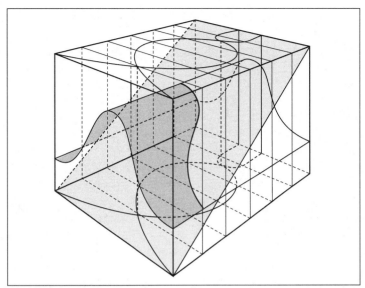

FIGURE 9: COMPLETE MIND MODEL

and will be explained in Chapter 4. These two natural growth processes work universally: in biological systems, in mechanical systems, in human relations, and in personal and collective development.

Take the example of Jonas Salk. Salk got so that he became "at one" with the polio virus, so that he could feel its intentionality and, from that intentionality, could develop the vaccine which was to become the cure for polio. He writes:

> . . . when I became a scientist, I would picture myself as a virus, or as a cancer cell, for example, and try to sense what it would be like to be either. I would also imagine myself as the immune system, and I would try to reconstruct what I would do as an immune system engaged in combating a virus or a cancer cell.

. . . I then discovered what the words identification and empathy meant. I could identify with a virus or a cell or with the immune system; I could also empathize with people in different situations, circumstances and states of mind. I realized that, objectively speaking, there was not one fixed point of view or one perspective, but many changing points of view that depended upon circumstances. . . . I developed the ability to shift my perspectives.

In this way I could manage to solve problems more easily. . . . I found myself at one with the object, or with the subject, and I could even project myself in time, through my imagination, and bring to realization intentions or imaginings as if they had become self-fulfilling prophecies. . . . I recognized the importance and value of the mind and the value of the game of empathetically shifting perspective in dealing with human problems as well as in unraveling the mysteries of nature. If the human mind can do one, it should be able to do the other.*

Or take the example of Nicola Tesla. Tesla invented over a thousand inventions that laid the foundation for our electronic age. He invented such things as transformers, amplifiers, switchboards, electric motors, and alternating current. And he did this by using his I.S.P. His process of Intuitive Sensory Perception was so acute he could "see" the potential of any invention vividly, as though it were already real, and he could sense it with all his senses. So when he imagined an electric motor, for instance, he could see that motor in every detail, he could take it apart, put it back together, and even watch it running so that the brushes wore down its copper cylinders—all in his mind. Tesla said it was both a blessing and a curse: a blessing because he could imagine an invention in such detail, but a curse because the image wouldn't go away until he actually built it.

*From Jonas Salk, *Anatomy of Reality* (New York: Columbia University Press, 1983), pp. 8-11.

When Einstein developed his theory of relativity, he announced his discovery to the world from a strictly rational, scientific perspective. It was not until some fifty years later that he admitted, in his personal journal, how he had come upon his amazing theory. He discovered the relationship between light, velocity, and mass by "becoming" a lightwave. He rode along with the other lightwaves and talked to them, and asked them how they worked. And that's how he conceived his theory of relativity.

Most great discoveries have been made by people using their Intuitive Sensory Perception. We all have within us a marvelous computer, our intuitive mind, that can take the whole context of things and, through its intuitive thinking, create ingenious solutions to our personal problems and for the betterment of humanity. We have now come to a time when every one of us—every worker, parent, businessperson, teacher, student, doctor, lawyer, and politician—must learn how to use intuitive thinking, learn how to do what all potentialized people have known how to do.

When Napoleon Hill interviewed five hundred of the world's richest, most successful people to find out the secret of their success, he discovered that they all used a similar process. They had an intuitive counsel, a set of intuitive guides, they would consult with whenever they were going to do a business deal, and before they made any decisions. They were using *Intuitive Sensory Perception,* and their guides were *mature holodynes!* This is one of our greatest secrets and it's so simple. Success means using I.S.P., accessing the holodynes on the holodynamic plane, making sure they are mature, and listening to their counsel, so you can potentialize any given situation.

How can you apply this in your daily life? You can learn to "track." "Tracking" is a simple, effective way to access your holodynes and take them up through the natural order of their own growth. Tracking is an internal process by which you can unfold the fullest potential of your mind.

3

Tracking

Behind every problem is a holodyne waiting to be tracked.

In the days of nomad hunters, "tracking" meant following the tracks of the animals in order to hunt them down. In the days of railroaders, "trackers" were those who laid the track for the railway. In holodynamics, "tracking" refers to the ability to track through certain problems to their solution. Tracking is the process by which a person who wants to get from point A to point B locates the blocks to his or her progress, and *potentializes* them. Before you can begin to track, you must own that the potential to get to point B is worthy of your focus, that the blocks exist—that there is a problem, and that, by using your I.S.P., you can personally help solve this problem. Then you are ready to track.

Tracking is the simple, but very powerful, process by which you access holodynes and then keep them "on track" through the natural stages of their development—mature them to their fullest potential. Early childhood experiences create memories which set up defenses within you, so you see only *part* of the picture. You get "off track" from your fullest potential. Because of these experiences, and certain family and cultural beliefs, part of your mind has never grown up. Tracking is a process of getting your holodynes "back on track," so they can complete their natural growth. Through tracking, you can take any immature thought or feeling and transform it into its fullest potential. Tracking uses all the dimensions of the Mind Model and has six steps, outlined in "The Tracker's Guide" below.

Note that the steps of tracking follow the stages of development—the natural Implicate Order of growth, as shown on the Mind Model. Note also that tracking is an *intuitive* process: you access your holodynes with your *intuitive* mind, and you keep them "on track" by focusing upon each step with your *intuitive* senses. Don't worry about "doing it right." Tracking is fail-free: if it turns rational, it will stop working, but no damage can be done. The process is completely under your control: to get "back on track," you simply return to the first step and start over again.

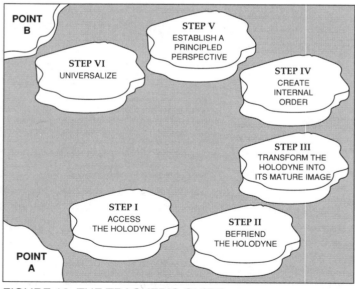

FIGURE 10: THE TRACKER'S GUIDE

Think of it as floating down an intuitive stream. If you get caught in a rational eddy, just get yourself back into the mainstream, and the whole process will continue right along.

You can track alone, with a partner, or with two or more partners. Alone, you will find it easier to get "off track": you are often so close to your problems that you cannot phase-space them. Having a good tracker as your partner will help you *stay* "on track," even when you get into downdraft dynamics you don't understand or don't perceive as limiting. It may be a good idea, at least at first, to have *two* partners—one to track and the other to help the *tracker* stay "on track." Each of these ways works, but tracking with *two* partners— a tracker and a helper—seems to work best overall.

The choice of staying in the process rests entirely with the individual being tracked. Part of your mind may wish to solve a problem and another part may resist solving it, but the choice is always up to you. Once you decide to track a problem, the six steps of tracking will guide you all the way through.

The Six Steps of Tracking

STEP ONE: ACCESS THE HOLODYNE

In order to access a holodyne, you must first: *focus* upon what you want; *own* that there is a problem, something blocking you from getting what you want; *choose* to solve the problem; *choose to use your I.S.P.* to solve the problem; *phase-space* for the solution; and then *select an internal guide* to help you. You will automatically do these things as you prepare to track. Once you are ready to solve a problem, accessing the world of holodynes can easily begin by *re-experiencing the feelings* associated with the problem. If you are stuck at point A when you really want to get to point B, here are some suggestions as you take your first step and access your holodynes.

FOCUS ON WHAT YOU WANT

In order to impact the Quantum Wave, you must *focus clearly* on what you want. You must clearly imagine, with all your senses, both rational and intuitive, what it would be like to *already* have what you want. In this way you can see where you are, compared to where you want to be, and begin to specifically locate the holodynes which are blocking you.

OWN THE PROBLEM

Before you can solve any problem, you, the owner, who created it and hold the keys to its solution, must step forward and *claim* it. Whatever you are experiencing, you created. To change your experience, you must create a new one. So by owning the problem, you begin the process of creating a solution.

Questions the tracker can ask:

"What is the problem?"

"What is your part of the problem?"

"How does the situation make you feel?"

"Can you own this feeling as your creation?"

"Would you like to solve your part of this problem?"

Once you *own* your part in the creation and the continuation of the problem, you need to *phase-space* the problem so you can locate solutions to the problem.

Phase-spacing for the Solution

When you get too bound up in the dynamics of your problem, you cannot free your mind to see the solution. Since your mind created the solution before it did the problem, you must clear your mind and phase-space the problem so that the solution can unfold. One of the best ways to do this is to use your Intuitive Sensory Perception and begin at your *place of peace.*

Suggestions by the tracker:

"Relax and, using all your senses, go to your place of peace. See the colors in their vividness. Hear the sounds of your place of peace. Smell the smells, taste the tastes, feel the feelings, and be that state of peace."

Now, if you've never had a place of peace and you were born in chaos and raised in chaos, and you cannot even imagine a "place of peace," just relax anyway. If you can think of *anywhere* peaceful—a beach, a meadow, a mountaintop, you can think of a place of peace. So just pretend. What would your place of peace be like? What would it feel like to be in a place where it was really peaceful? And then use all your senses to imagine that place, to feel it, and to be its peace.

When you are fully relaxed, and your mind is at peace, you are ready to find your intuitive guide.

USE AN INTUITIVE GUIDE

Since most problems require complex, and therefore *intuitive,* mental processes, and since the solution to any problem is enfolded within your mind, it is essential that you call forth a guide who can *intuitively* help you solve your particular problem. Usually your Full Potential Self—your "I"—or some other mature guide will appear in your mind to help you through the problem.

Questions the tracker can ask:

"Do you have a guide who will help you with the solution to this problem?"
"Who (or what) is your guide?"

"Can you call forth this guide?"

Now don't worry if you can't "see" your guide, because it can appear to you through *any* of your six intuitive senses. But make sure you find a way to communicate with it. Once your guide has appeared, ask if it will help with the solution to your problem. If the guide says yes, or indicates agreement through some sensory channel, go right into the feelings that are associated with the problem. If the guide says no, you may want to explore a little. Ask the guide what its problem is. If it will not tell you, then look for another, until you find one that you really feel good about and that will communicate with you.

RE-EXPERIENCE THE FEELINGS
ASSOCIATED WITH THE PROBLEM

Re-experience the conditions and situations that created the problem, or that allowed it to surface. If you have a problem with losing your temper, think of situations that cause you to lose your temper, such as a driver pulling out in front of you on the freeway and then slowing down. Go into the exact feelings you have, as you're pounding on your dashboard, cursing the other driver, if that's what you do, get into those feelings and feel them completely. Re-experience the problem in detail, so that all of your senses are involved.

Re-experiencing the problem activates or re-creates the resonating energy field of the holodynes involved. All the resonating frequencies from the holodynes become clear, so that you can travel down these frequencies intuitively, and access the specific holodyne which is directly creating the anger reaction.

Suggestions by the tracker:

"See yourself experiencing the problem. Relive the experience as best you can. Recall all the feelings associated with
it and accept fully your part in it. What are you feeling?"

Once you have entered the resonating energy field of the
problem, use your Intuitive Sensory Perception to sense its physical dimensions.

GIVE THE HOLODYNE A PHYSICAL MANIFESTATION

In order to access the holodyne creating your problem, you
must switch to your intuitive mind and give the feelings associated with the problem some physical form. This meets the boundary conditions of the first stage of development — the first level of
the Implicate Order.

You may use any or all of your I.S.P. channels, your intuitive
senses. See the form, feel its texture, taste it, smell it, or "sense" its
presence. Use your intuitive guide to help you create the form. It's
easy and natural because it reflects an actual image already operating in your mind, on your holodynamic plane. By re-experiencing the problem, you are calling forth into *conscious* awareness the
holodyne that has created the problem, and your intuitive mind
will work it right every time.

Questions the tracker can ask:

"Can you go into the feeling?"
"What does the feeling look like?"
"Does it have a shape?"
"Does it have a color?"

If one Intuitive Sensory Perception channel is not open, use some other channel.

The tracker might ask:

"Is there a presence you can sense?"

"Can you hear (smell, taste, or feel) it?"

"Does this feeling store itself somewhere in your body?"

"Where?"

Tension or stress for instance, will sometimes store itself in the intestine or lower back. So ask yourself, "Can I pull the feeling out, put it in my hand, and see if it has a shape or a smell or some sensory quality that it will allow to unfold?" Be aware of and respect your intuitive feelings. Do not allow your rational mind to impose artificial images upon your intuitive mind. Always allow your intuitive guides to direct the process, encourage them. When an image is presented, accept it immediately, no matter how repulsive it may seem at first.

If there are too many images, or no image appears, ask your guides to help you. Store extra images until you can deal with them. Use your imagination to help you create other images. Don't worry if you don't actually "see" an image. Any sensory channel will do. But it helps to have *all* sensory channels in operation when you track a holodyne: you get a better picture when all the lenses in your mind camera are opened. Practice. It comes easy if you relax and let your I.S.P. do the work.

Once you have accessed the holodyne and found a shape or a presence you can sense, you have completed the first step in the Implicate Order of tracking. You have given the holodyne a physical manifestation. And you are now ready for the second step.

STEP TWO: BEFRIEND THE HOLODYNE

The second step is to unconditionally accept the holodyne as a friend. In the Implicate Order of life, the second stage of development reflects self-awareness. Some of these holodynes will be so different from anything you have ever experienced that you will have a tendency to reject them or be repulsed by them. Remember that holodynes come from the quantum dimension and operate on a different order from our own. They can take the form of anything from a tornado to a burning sun, to a black blob, to *anything* you can imagine. They do not seem to have the same environmental restrictions on shape that we do. Nevertheless, they do manifest themselves in specific shapes.

By unconditionally accepting the holodyne, you allow it to manifest its "self." If you treat it like a friend, it will respond like a friend. You can ask it questions, and it will give you answers. However you treat the holodyne, it will respond in kind. You cannot fool a holodyne because it is holodynamic and therefore knows more about you than you consciously know about yourself. It knows how you are really feeling, whether you admit it or not. If you are repulsed by the holodyne, or in any way seek to repress or reject it, it will create all kinds of mind games to keep you struggling.

The list of mischief and evil that holodynes can do in your life is almost endless. Holodynes are in charge of all your illnesses, broken relationships, failures, pain, misery, fear, guilt, resentment, and inadequacies. Turbulence and chaos are all created by holodynes. Even though you've spent all of your life rationally fighting these subconscious saboteurs, it is time now to use a different approach. It is time now to go deeper and use love, which is the most powerful force in the universe. Once you accept the holodyne with unconditional love, it will tell you all about what it has been doing, and what it wants.

Questions the tracker can ask:

"Will it communicate with you?"

"What has it been doing in you?"

"Can you thank it for all it has done?"

Thanking the holodyne is vital. These thought-forms are very sensitive and will drop back into your subconscious mind if they are resisted in any way. Remember that an atmosphere of unconditional, positive acceptance is essential. Once you have become friends with the holodyne, you are in a position to identify its intent.

FIND THE HOLODYNE'S POSITIVE INTENT

Ask the immature holodyne what it wants. Help it focus so you can discern its intent. By finding the holodyne's positive intent, you are letting it achieve its potential in a more mature way: the holodyne which once functioned as a block in your mind can now become a helper. Allow your unconditional love to focus upon its positive intent.

Questions the tracker can ask:

"What is it you really want?"

"If you continued your activities what would you eventually get?"

"What would it feel like to have what you really want?"

Usually the holodyne will come right out and tell you what it wants. Sometimes this will be very negative. It might say something like: "I hate you," or "I want to kill you." Don't ever worry about the intent of a holodyne, no matter how negative it first appears. You will always find its positive, real intent if you keep exploring. When you confront a primitive holodyne that doesn't know its positive, real intent, simply ask it, "Well, if you hate me (or kill me, or whatever), what will you *finally* get from that?" And it may come back with "Peace," which is what it really wants. That's the problem with immature holodynes: they may really want the best of things, but their way of getting them is immature. Once you find out the holodyne's positive intent, thank it, remain friends, and continue on to Step Three of the tracking.

STEP THREE: TRANSFORM THE HOLODYNE
INTO ITS MATURE IMAGE

Return to your place of peace and relax. You are going to *transform* the immature holodyne into its mature potential. To do this, you need to take the holodyne's positive intent and, using your I.S.P., see what it would look like in the shape of a mature holodyne. The positive intent reflects the enfolded Full Potential Self, the "I," of the immature holodyne. Using I.S.P., you can call forth the full potential of the immature holodyne and accelerate its natural unfolding.

Suggestions by the tracker:

"Ask your guide to show you a new, mature image of the holodyne which can accomplish the holodyne's positive intent in a mature manner. Use your I.S.P. to sense fully this new holodyne."

If you have a mature image of the holodyne, continue on to the next set of questions. If you do not, switch to some other sensory channel. Explore until you are able to get a mature image.

Questions the tracker can ask:

"Will the holodyne's mature image help you accomplish the positive intent of the immature holodyne?"

If it will, invite it into the presence of the immature holodyne and ask:

"Can the mature holodyne and the immature one *come into each other's presence?*"

"Can the mature one *thank the immature one* for all that it has done?"

"Can the mature one *demonstrate that it has the power to accomplish the positive intent* of the immature one?"

"Is the immature one willing to allow itself to grow up and to be transformed into the mature one at this time?"

If the immature holodyne is willing, *let it be transformed.* This is a natural process. It's spontaneous. You will find that the new holodyne will simply absorb the immature one right into itself. You can actually feel the transformation; it changes the flow of life energy through your body, as well as through your mind, in ways you can pick up with your senses.

If the immature holodyne, or some part of it, is not willing to be absorbed, you will know it immediately. An unabsorbed part often indicates a separate dynamic, which you must treat individually—as a separate holodyne with its own issues. Explore, ask it *why* it is unwilling to be absorbed. Usually there is some aspect of the immature holodyne which you have not fully exposed and which you need to understand before going on. If you treat each unabsorbed part with respect, as a "self," and find its positive intent, it will be absorbed—transformed—into a mature image.

Once the transformation is completed, *commit* the mature holodyne to "be there" for you, to develop real connectedness, and to maintain open communication with you from now on.

Questions the tracker can ask:

"Will you (the mature holodyne) be there for this person from now on?"

"Will you (the person) commit to be there for your mature holodyne from now on?"

"Will you *both* commit to real, open communication with each other?"

Once this is completed, go on to the next step of tracking.

Remember, your intuitive mind is always truthful. It will not present a holodyne as mature until all the issues related to its purpose are fulfilled. So the third step is where the *real* work of tracking—and of personal development—is done. Pay special attention and *make sure you complete this step* before proceeding to Step Four.

STEP FOUR: CREATE INTERNAL ORDER

It is not enough that immature holodynes are transformed into mature ones. In order to achieve your full potential, you must be willing to establish in your mind an internal system of government similar to the external one you would like to see established in world outside you. In the Implicate Order of life, your new set of mature holodynes must become a "living" system, with its own causal potency, its own mature "field of influence." Once you establish a system of government in your inner world, among your holodynes, you become more effective in the outer world, among people.

Referring back to the Mind Model allows you to establish among your mature holodynes a system of social order, of friendship, trust, and teamwork. This order is necessary because only with it can your mature holodynes become aligned with each other and with your Full Potential Self, and so tap into the Quantum Force. Step Four creates order and aligns all holodynes on the holodynamic plane so they can combine efforts to solve any problem and see that it *stays* solved. Without this alignment, the immature holodynes cannot be disciplined and the same old problem arises, over and over again. The crucial choice is whether to *act*. Will the mature holodynes act on their new commitment to cooperate and communicate, to be there for you, to help you find the solution to your problem, or not?

Questions the tracker can ask:

"Are you willing to teach everything this person needs to know in order to solve this problem?"

"Are you willing to dedicate everything you are to this person's fullest potential?"

Once dedicated, the mature holodyne has made clear that its deepest intentions are for the unfolding of your potential. This sets up a special resonating energy field which aligns the holodynes on the holodynamic plane of your mind with each other and with your "I." And this alignment clears the way for the Quantum Force to flow freely through the holodynamic plane and for your fullest potential to unfold. So, if the mature holodyne is willing to dedicate itself, go on to the next step.

STEP FIVE: ESTABLISH A PRINCIPLED PERSPECTIVE

In this step, you clarify and put into action the guiding principles by which you and the mature holodynes of your mind will operate. If you want a world that is caring, fair, and sharing, you need to "become" caring, fairness, and sharing and, through your mature holodynes, to set up a resonating energy field which will demonstrate and realize these principles.

Questions the tracker can ask:

"Will you care, be fair, and share?"

"Will you treat all others as you wish to be treated?"

"Will you operate on the basis of equal rights?"

"Are there any other principles you would like to establish as part of the internal social order?"

When you ask the mature holodyne not just to be open, communicative, and cooperative, to be there for you, and to make every effort on your behalf, but also to *"become"* your guiding principles—to *"become"* fairness, caring, and sharing, you set the stage for the manifestation of a higher order, in which all parts of your inner self, the "I," and your conscious self, the you-in-this-world, can come together in a new, integrated way.

This "coming together"—this principled connecting of the "I" with the "you"—can transform your personality so dramatically it is sometimes called "the quantum leap." It is as though the logjam in the stream of your life suddenly let go and all your energies started flowing again. It is like coming home. You are at peace. You are love itself. You don't really need anything or anyone, but you are more connected than ever with everyone and everything. You are more able and willing to reach out to others in such a way that everyone and everything is "updrafted."

Step Five brings you a deep sense of integrity within your mind and throughout your being. Whatever you do, you can honestly say, "I *live* what I am." You have owned all your holodynes and you can see their impact upon your life. You have *owned* your life and you can see how you caused every circumstance you have ever experienced. All the old "victim-blamer-rescuing" games are transformed. You are now ready for the next and final step of tracking.

STEP SIX: UNIVERSALIZE

You have tracked your mature holodyne through to its fullest potential. In this step, you *invite the mature holodyne to meet with you on a regular basis.* At least twice a day, using your full I.S.P. powers, convene with it in your place of peace. Many find it convenient to create a special conference table, usually a round table, so that everyone—your conscious self, your "I," your intuitive guides, and other mature holodynes—can meet on an equal basis. Because holodynes have the power to cause things to happen and, if treated right, to cause the *right* things to happen, you can begin immediately to show trust for the new holodyne and to affirm its value by *giving the mature holodyne a specific assignment.* The most appropriate assignment would likely deal with the problem you have just tracked. Ask it to help solve the problem. Ask what *you* can do in order to help. Ask it to report back at a given time. This arrangement creates a real exchange of information and keeps everyone up on assignments; it gives your rational mind open access to your intuitive processes—particularly helpful in times of stress or major blockages.

Once your mature holodyne is situated at your round table in your place of peace, you have available a new source of intuitive information and strength. You have opened a new channel between your intuitive source, the Quantum Wave, and your rational mind. You can now access the fullest potential of the holodynes involved in any problem and create new solutions to that problem. Now the real fun begins!

You can now extend your mind and your influence into wider and wider spheres. Attuned as it is to the holodynamic plane within, your mind can now become attuned to the holodynamic plane of your external world. You can reach out in a loving, knowing way and, by the reaching out, fulfill more and more of your own potential. You come to experience your oneness with other people and with our great, living, holodynamic universe. You universalize.

At this point, you have completed the tracking process for a specific holodyne. You have brought it from its hidden, immature form to the open, mature image of its Full Potential Self. You have aligned the individual holodyne with your "I" and with other mature holodynes on your holodynamic plane. You have established a system of government among your holodynes and, within that system, you have aligned each holodyne with certain guiding principles. You have given your mature holodynes a way to manifest these principles to you by having regular meetings, working on projects, and celebrating life. And thus, you have aligned your conscious self with the formerly subconscious dynamics of your mind.

There are some powerful things that you can do within the conference setting of your round table. You can outline specific problems you want your mature holodynes to help you with, and behind the scenes they will set up resonating energy fields that will cultivate solutions without your *consciously* having to put out any really hard work. They will make things easy. They will make things natural and spontaneous. Just as a seed will grow if you water it and take care of it, so if you maintain your conscious focus and stay aligned with your fullest potential and with the mature holodynes of your round table, your solutions will blossom!

When you give specific assignments to your holodynes, the more you focus upon those assignments, and the more loving energy you put into your holodynes, the greater their positive causal potency. When you combine your conscious, rational focus with your caring, intuitive focus, your holodynes respond; you can work together as a team. Follow through on their assignments with regular "reality checks" to make sure those assignments are done, but remember that holodynes operate on a different order of time from ours. Your rational, linear mind may often expect them to fulfill their assignments by *its* time, but the holodynamic plane is a developmental plane and there all things grow according to a natural order. Once you're tuned in to that, you can balance the demands of your rational mind with the reality of your intuitive mind by setting realistic time frames, so your rational mind won't get nervous if its preconceived schedule is not met.

You have within you a galaxy of wonders. You are now free to explore whatever dimensions of your inner world you would like. Within you are all your experiences of this life. By accessing holodynes on your holodynamic plane and maturing them through the natural stages of their own growth, you can remove all your emotional blocks and free all the resources of your heart.

When your mature holodynes are aligned with your Full Potential Self, you can tap more deeply into the power behind all reality, the Quantum Force, and use it for your life: you can draw boundless life energy from its very source. Vitality, abundance, and health are yours for the taking.

Using your Intuitive Sensory Perception, you can turn your senses outward to other people and develop universal-mindedness towards every person and every life-form. You can access their Full Potential Selves, carry on dialogues with them in your mind, and, through the resonating energy fields of their potentials, you can help them unfold their enfolded potential. This is one of the greatest manifestations of universal love possible.

Case Histories

Tracking can help people of all ages. One evening, I was in a group discussion at a friend's home. An eight-year-old girl was sitting beside me, listening intently. I noticed her face begin to look strange. It was swelling rapidly. I asked her what the problem was. Apologetically, she answered, "There must be a cat in the room."

We found the cat and I suggested it be left in the room. Since we had been talking about tracking, we could experiment to see what would happen if we tracked the little girl's allergy reactions to cats. She enthusiastically agreed. I helped her to relax, to go to a place of peace and see if she could call forth the allergy.

Almost immediately, she said, "It's a fuzz-ball!" It was brown and it lived in her nose.

"Can you make friends with it?" I asked.

"Oh, yes," she said. "She is warning me about cats. She doesn't like cats."

"How come she doesn't like cats?"

"Just because. She just doesn't like 'em."

"Tell your fuzz-ball you want to know more. Ask her when was the first time she didn't like them," I said.

She began to cry. "It was when I was three."

"What was happening when you were three?"

"My Daddy was gone helping people get better (her father was a doctor) and I wanted him to be with me, so I got sick."

"So the only way you could get your Daddy to love you when you were three was to have this allergy?" She agreed. Her father, sitting next to her, was heart-stricken by her answer.

We had her envision a mature image of the fuzz-ball for loving her father, and she came up with a beautiful yellow flower.

We introduced the flower to the fuzz-ball and a transformation immediately took place. The flower agreed to "be there" for her, to help her develop a better relationship with her father, so that, even though he was very busy with his patients, they could plan special times together. Her father reached out and took her in his arms.

The whole process took about five minutes. The swelling in her face disappeared, even though the cat was still in the room. As far as I know, her allergy has not returned. She and her father continue to do those "special things" together.

In cases like this, where psychological stress has lodged itself in bodily functions, tracking can have a dramatic and direct impact.

TRACKING A TEMPER TANTRUM

As Murray and Ethel entered the room, I was immediately impressed with Murray's size. He literally filled the office door. He offered his hand and said, "I have a terrible temper!" His wife, a five-foot-two twiggy of a person, peeked out from behind him, first on one side, then on the other, and said, "Oh, he does. Yes, he does."

"Oh, really?" I responded, as I met him eye to eye (I had to look up even though I'm just about six feet tall myself). I reached out and took his hand, as I said, "When did you get this terrible temper?" Taking note of his iron grip and his mental strength, I motioned to a couple of chairs. Murray paused. His face crunched up, he looked at me out of his left eye and said, "I guess I've had it all my life!"

"So how old are you when you're using it?" I asked matter-of-factly, still holding his hand. As he focused on "when" he first got his temper, I let go of his hand and he settled into his chair.

"What's the question again?" he asked, almost absently.

"The question is, when did you first learn to use your temper?

How old were you? And what's more important, how old are you when you use it now—*really?*"

"I guess I'm pretty young, aren't I?" he said with a kind of twinkle. I could feel his relief that he couldn't intimidate me with his anger, his power, or his temper. Already we had subtly established that it was all right to have a temper and that it would be interesting to look at tempers more or less objectively.

"Ever take a look at that temper tantrum habit?" I suggested, taking the lead.

"No, not really," he pondered. "What do you mean?"

"People usually get angry or throw temper tantrums because they want something they care about. Anger is just an immature way of caring. So when you are angry, it's because . . . ?" I let it dangle.

"Well, yeah. I guess it is because I care. I never thought about it that way before. Have you, Ethel?"

We were already getting to "where he lived." No stories—just directly, openly putting his anger on the table. I liked this man, and I didn't want Ethel to take over for him. Murray was big, but I had a hunch Ethel was big enough, in that little body of hers, to handle him most of the time. If she kept doing this for him, it would only prolong his problem. He had to face this, and own it, if he was ever going to move beyond it.

"Go back to when you first began to throw temper tantrums," I suggested. "Take a look. See if you can look at why your mind decided to develop the temper tantrum tactic. It's a habit you have evidently used a lot in your life, and so you must be pretty good at it."

"Oh, he is. Yes, indeed, he is!" Ethel chirped in. She had a delightful smile and bright blue, beaming eyes. She sat right up in her chair, even though it was a swivel rocker. She held her purse in her lap with both hands on top of it.

I shifted my gaze back to Murray. "Can you, in your imagination, envision the little boy who's so angry?" I asked. He looked up, closed his eyes, then hung his head. "What is it?" I asked gently. His wife was wide-eyed, staring at him, a little astonished that he would openly feel feelings she had seldom, in all her thirty-two years with him, seen him reveal.

"I am lying on the floor, screaming," he mused. "My mother is going to leave me with the baby sitter. I don't like that baby sitter. She's mean to me. She pinches me." He opened his eyes and looked right at me.

"Stay right with that little boy." I knew if he would allow himself to experience the trauma he felt then, to see himself through his adult eyes as he was then, he would be able to get a much clearer picture of the images which created and controlled his temper tantrums. He closed his eyes again.

"How old are you?"

"Why, I'm only two years old, or maybe three! I hated this baby sitter. My mother kept getting her and she didn't like me. I had no way to tell my mother, so I threw a fit."

I wanted him to feel, to explore, to relive as much of this as he wanted. Even though we hardly knew each other, a bond had formed between us. He stayed in the image, or holodyne, and felt what he had repressed all these years. He was reliving what he felt as a child. He had tapped into the source of his temper and had given manifestation to the holodyne involved. It was himself at age two or three.

"Just feel it all." Ethel and I waited quietly. He began to cry. Ethel moved over to comfort him. I held up my hand. "Just let him experience this in his own way. He'll be fine." She looked a little worried, a little motherly, but she settled back into her chair.

"What's happening?" I asked softly.

"The little boy feels like he's been let down, kind of betrayed. His mother doesn't care." He had tears coming down his cheeks. Ethel picked a tissue out of her purse and put it in his hand. He hardly seemed to notice.

"So, what did the little boy decide?"

"He decided that, to get people to care, he better get mad!" It was like a light coming on in his mind. He opened his eyes. He was grinning. He was getting to know this temper tantrum holodyne.

"Stay with it." I paused. I could feel his mind. He closed his eyes again. "Let's get to know this little boy. Talk to him. Make friends with him. Let's find out what he's like."

Murray kept his eyes closed for the duration of his "mind journey," relating in a warm and open way with this memory of himself. He was actually getting to know a part of his mind not usually accessible to his conscious self. He began to understand how desperately he had wanted people to care when he was just two or three. He wept at the love this little boy felt in the face of impossible odds.

"Let the little boy know how much you appreciate him and that you really *care*," I suggested. He began to rock back and forth.

"I am holding the little boy in my arms and telling him I love him. He's very relieved to know it and feels comfortable in my arms. I think this is the first time he's been comfortable for a long time."

After a few moments, I suggested we begin to explore how much influence this little boy had exerted throughout his life. We laughed together at how often he'd been angry, and I suggested he ask the little boy to help him understand how "helpful" he had been at different ages. Murray smiled a knowing kind of grin.

"What is it?"

"He's showing me a book. It's the book of my life! He's pointing to various times when he took over for me in order to get more caring. I can see the pages turn. I can see the boy. He's very active in my life."

"How is he affecting your life as you grow older?" I wanted him to look at how this early life experience may have set a pattern for his actions throughout his life, how it may have influenced his relationships with others.

"I'm six and . . ." he started to laugh. "The teacher won't let me play with the other kids in the ball circle. I'm too big and I have to sit out. I'm mad, so I won't go out of the circle. He has to drag me away. I'm real mad."

"What do you do?"

"I bit his leg! I'd forgotten all about that! I can't believe I bit the teacher's leg. Served him right though."

"Is your two-year-old there, in your mind? Can you see him now?"

"Yes. He's delighted. He's laughing. I think he wanted me to bite the teacher."

Ethel was holding her breath. She could scarcely believe that, in just a few minutes, we were in the middle of her husband's mind and tracking things she never knew about him. She sat there with her mouth open, wide-eyed and agape at what might come next.

We tracked the little boy in Murray's mind through four or five more situations—at fourteen, he got threatened and bluffed his way out of a gang confrontation by playing the angry tough guy—at twenty, he got rid of a bothersome friend by throwing a temper tantrum—at thirty, he lost an important deal because of his temper, and so on. In each case, the little two-year-old was there, orchestrating and supporting the whole action scenario. Finally, I suggested that we talk to the two-year-old about what it was he really wanted.

"Will he tell you?"

"Oh, sure. He says he just wants people to be fair and to care. He says he gets mad when they are not fair."

"Is there anything else he wants?"

"He says he wants me to be able to make friends and be loved." At this, Murray softened visibly. His huge frame let go of something and he slumped down into his chair, his shoulders sagging a bit, his hands folded into each other. Ethel couldn't stand it any longer. She went over and put her arm around him.

He was alone in his thoughts for a minute, then I asked, "Is the boy willing to really help you with fairness and love?"

"Oh, sure. He says that's what he's been doing!"

"Ask him to wait a moment."

"He says okay. He'll wait."

"Now, Murray, and you, too, Ethel, this little boy that Murray is talking to is a memory. I call this a holodyne. It's a whole living memory inside of your head. Everyone has them. Almost everything we do is controlled by them. This little boy has never had a chance to grow up. He's still trying to get you fairness, love, and friendship by throwing temper tantrums. Now, I ask you, Murray, does throwing temper tantrums *really* get you friends or *really* get you love and fairness?"

"No, of course not." His adult frame of mind could see that when he was acting like this little child and throwing temper tantrums, he paid a price in his adult world. He could see how his long-term investment in temper tantrums had kept him from friendship and thwarted real intimacy with his wife and with others.

"Do you think there might be a better way to achieve these things? A more mature way?"

"I would hope so, Doc. That's why we came to you."

I pulled out a diagram of the Mind Model. "This is a model or a map of how your mind naturally works. This area here shows the six natural stages your mind usually goes through as it matures. When you have a memory, or a holodyne, like your little two-year-old boy, Murray, that part of your mind has never had a chance to go through its natural stages of development. Somehow it got locked in at two years old and has never grown up. Do you know anyone else that acts like a two-year-old every now and then?"

They both smiled. "Yeah. Don't we all?"

"Let's help your holodyne grow up, shall we?" They both sat right up. "I want you to picture a new image in your mind. Recall, if you can, Murray, some person in your life who really knew how to love, make friends, and be fair." He paused, closed his eyes, and we waited about ten seconds. I could see his rapid eye movements beneath his eyelids. Then his expression changed. He seemed to relax.

"It's my grampa," he volunteered, a little awestruck. "He was a wonderful man. *He is radiant!*" he marvelled. I gave him time to more completely experience the memories he had of his grandfather, and how loving, fair, and friendly he was, as a being of light. In this way, Murray was able to access the image, the mature holodyne of his grandfather, very specifically.

"Will he help you with this? Ask the image of your grampa."

"Yes. He says he will."

"Okay. Now, Murray, we are going to help your two-year-old fulfill his purpose in your mind. He can only do this if he gives up his old patterns and adopts some new ones. Would you like to continue with your temper tantrums or would you rather relate the way your grampa related?"

"Well, of course. I'd love to be able to handle things as well as my grampa did. He was a great man!"

"Okay, then your little boy must allow himself to grow up. Do you think that if he really grew up he would become like Grampa?"

"Well, now that I think about it, I guess he would!"

"If you'll allow him to grow up, into the very image of your grampa, you'll release him from his own trap and free the energy he is using for temper tantrums so that it can be used for things which Grampa would do. In this way, your immature two-year-old will be absorbed into your grampa, so that whenever you go to throw a temper tantrum, you will act like Grampa instead. Would that be okay?"

"Yes, but my little two-year-old is nervous. He doesn't know what that means."

"Tell him everything will be all right. This is his big chance to help you get the fairness, the love, and the friendship you have always wanted. Ask him if he is willing to take this next step in order to help you get what he has always wanted for you."

"He says he is. Yes. It's, it's like that's *all he ever wanted from the beginning!*"

"That's what we find, time after time, Murray. All those times you blew your stack were orchestrated by this little two- year-old. All this part of your mind, this little holodyne, ever wanted was fairness, friendship, and love. So let's help him get what he wants, shall we?"

Murray nodded his agreement. He was ready. His mind was attuned to this new possibility. Never had he been so close to the source of his perpetual anger. Never had he grasped the deep intent of all his temper tantrums, his insecurities about friendship, love, and fairness with such depth and clarity. His eyes were still closed.

He was exploring a world within himself, a world which only revealed itself indirectly, a world which had awesome power in his life. Even more exciting, he had just received a message that he could take charge and create what he wanted in this inner world. He could tap right into his own source.

"Have your grampa come into your two-year-old's presence, take him on his knee, and love him for all the things he has done for you. Let him feel your love and Grampa's love. Now explain that the next step is for him to be absorbed into Grampa so that all his power may be used to love maturely as Grampa would. When he is ready, just let him, in your imagination, be absorbed into Grampa."

"It's done. I feel so light. I feel great. Wow!"

"Okay. Just a few more steps. If you look at the Mind Model, you can see that, when you located the little boy in your mind, you were actually at Stage One. As the little boy took on a personality and talked with you, you had progressed to Stage Two. When you found his real intent and then gave that real intent a mature image, Grampa, and had Grampa absorb the little boy, you were at Stage Three. Now, ask Grampa if he will be there for you, committed to teach you all you need to learn about love, friendship, and fairness, for the rest of your days. Will he?

"He says he would be glad to. Yes."

"Good. You have now established discipline and teamwork in your mind, so you have progressed through Stage Four. Now ask Grampa if he will allow you access to him anytime you like, and ask him if he will serve in your mind, based upon the universal principles of unconditional love, peace, fairness, and sharing."

"Yes. He will."

"Okay. Now, ask him if he will sit at a conference table in a special place of peace where you and he can always find each other. You may want to give him some assignment. Is there anything specific you would like him to handle? Any place where your temper tantrums have been getting in your way?"

"Well, yes. There is. Ethel and I have been having our differences and my temper has been scaring her of late. I think she's getting tired of it."

"Ask Grampa about your temper. Why has it been in your life?"

He closed his eyes, bowed his head a bit, and contemplated. "Why, he says it is *my adventure with love*. Every temper tantrum showed me less than real love. It was a way to help me appreciate real love." He wept. So did Ethel.

We then proceeded to talk about how the anger had functioned in their relationship. It didn't take Murray but a moment to see how each angry outburst was an immature attempt to obtain a deeper loving interaction with Ethel. The conversation naturally turned to what Ethel's part in the "anger dance" had been.

"You mean to say that *I* had a part in him getting angry?" she exclaimed, rather befuddled at the idea. "I can't see how I had anything to do with it." She said it with such finality and such dignity that I felt as though I had just been dismissed by a queen.

"Ethel," I said softly, "can you just entertain the thought that maybe, just maybe, you've been a victim by your own choice?" The question hung in the air. This was a powerful little lady. I just sat there. She turned and looked out the window. I continued, "Have you ever considered that anger and other forms of emotional abuse require not only an attacker but a victim and a rescuer? If Murray has been angry with you to the point that you have come in for help, do you want to look at the part of you that feels like the victim?"

"You're not going to blame his anger on *me*. I have *nothing* to do with it." She was cold and factual. Her mind had shut off like throwing a switch on a light. I turned to Murray.

"What are you feeling, Murray?"

"I don't know. I'm sort of shut down."

"Look a little closer. What is your body telling you?"

"I feel upset. I guess I'm frustrated. I don't think you should push this onto her. She's not to blame for my anger or my abuse of her. I am. I realize that and I want to change."

"Murray, the process you are using right now protects Ethel and doesn't solve the real problem, because protecting Ethel is the other side of abusing her. Are you willing to ask your mature Grampa image what he would suggest?"

"Oh, sure." He closed his eyes. After a moment he reported: "Grampa says that Ethel wants to be sure that I don't go back to the abuse and the temper. She's afraid that if she takes responsibility for my anger, I won't change."

"Ask Grampa what you can do to help."

"He says just love and accept her. She doesn't need any help. She just wants a little time to see if this process really works." At each point that some new pattern was needed, I would suggest that Murray ask Grampa what he would suggest be done. Grampa became his guide for the rest of the session.

At first, Ethel did not want to deal with her part of the problem.

Rather than continue to confront her fear directly, after she had seen Murray "in action" a few times, I asked her if she had a special place of peace. She replied that she did. It took a bit of encouragement for her to switch from her rational control to her intuitive mind, but at last she went there briefly and felt the peace. It was all she would allow herself during this first session.

At the end of the session, I took Ethel's hand in both of mine. She looked at me with clear, blue eyes filled with relief, trust, and appreciation. "Ethel," I said. "I want you to do something before our next session. I want you to take this page with the six stages of development (see Chart 1, p. 48) and I want you to study it. Murray, will you do this with her?" He nodded that he would. "Get so that you both know each stage. Discuss it together. Then see if you can both have some fun locating images, or holodynes, that perhaps have been influencing each of you at each stage. Next time we'll take another look. And one more thing. Most of the holodynes which govern our lives are modeled from our family tree. So here is an outline of how to look at your family tree, your parents, grandparents, and others who have been your models for all your holodynes. I'd like you both to fill out this genogram (see Chart 2, p. 208) as best you can, so we can discuss it next time. Okay?"

She leaned forward and gave me a bit of a hug. "Whatever you say, Doc," she smiled. Murray came over and took my hand as we moved toward the door. "Thanks, Doc. That was great! I never felt better. See you next week." He couldn't get through the door while still holding my hand. "Oh, sorry," he said meekly with a childlike smile. He let go and moved sideways through the door. I waved before I turned back.

What Murray and Ethel learned in the first few minutes of tracking was that Murray's mind had been orchestrating circumstance after circumstance (temper tantrum after temper tantrum) in order to come to grips with fairness, friendship, and love. A part of his mind (the two-year-old) wanted the fairness, friendship, and love but did not know how to get it, except by throwing temper tantrums. This part of his mind was waiting for some new development which would unlock it from its childhood trauma.

Ethel was part of the problem. As the victim, she was highly controlling and rationally defensive, with a whole series of expectations about Murray, his life, their relationship, and how things "should" be. None of this had any impact on his temper tantrums, except to throw fuel on the already smoldering holodynes. Ethel and Murray set up a resonating energy field within their relationship: her rational control pitted against his irrational temper. Their marriage became a great, entertaining, painfully real battleground for both of them.

Ethel refused to give up her "rational controller" at first. She needed a little time to see if things were going to change, and she wanted to see if she could "pick away" at Murray's new, mature holodyne. It was a way of continuing the game. It was also a way of ensuring that change was sincere and effective on Murray's part. In the next session she confessed that Murray had not had any temper tantrums, in spite of her efforts to give him every opportunity. He chose to empower "Grampa" in each instance. So Ethel was ready for her *own* tracking by the time she got to the next session.

In later sessions, we explored the brutality of Murray's anger flashes and their effect on Ethel. Murray had emotionally and physically abused Ethel, and several others. Her fear of him had closed her off, and shut down their sexual intimacy, their spontaneity, and, as Ethel described in later sessions, "was slowly killing our relationship."

As we addressed Murray's anger successfully and he was able to mature beyond it into adult caring, we could then access Ethel's fears and shut-down patterns. She already knew these were defensive reactions. The exciting part for her, and for all of us, was to experience her transformation as she matured along with Murray. She learned to track his emotional intensity and to facilitate his development, rather than to become rationally defensive or controlling or mothering, and therefore emotionally shut down. A new dimension was created in their relationship.

Most of us are like Murray and Ethel: early in life we develop the immature holodynes which orchestrate our problems—both individual and collective. What we call "social ills," such as crime, drug abuse, alcoholism, poverty, and ignorance, and what we call "world problems," such as war, famine, economic depression, and pollution, are all the result of such immature holodynes, the holodynes behind our primitive collective beliefs. Whenever we allow these beliefs to go unchallenged, we empower their immature holodynes and so become part of the problems, rather than part of their solution.

A universal solution rests in tracking the immature holodynes behind these problems through to maturity. How much anger in the world could be overcome and matured into caring and friendship if people everywhere were to experience what Ethel and Murray did. What impact we could all have if people everywhere were to learn the principles of holodynamics and apply the specific skills of tracking to mature the holodynes causing primitive feelings like anger.

Can we *really* change the physics of our minds? Can we consciously *will* ourselves to live in a better way? Can this rising tide of crime, murder, rape, incest, drug addiction, alcoholism, abuse of self, and abuse of others *really* be turned? Is it possible for each of us to have a real impact on our world? My answer, from long experience, is an emphatic "Yes!"

Every day I get letters and phone calls from people amazed that *the process works,* amazed at the dramatic difference tracking has made in their lives. Some have achieved, in moments, the peace they had always sought. Others have found deeper meaning in their intimate relationships. Families have become more unified, marriages have become more fulfilling, personal lives have found a new quality of integrity and harmony.

Here is a third case history. This one deals with intimacy.

TRACKING INTIMACY

A young couple wanted to see me on their lunch break. I reviewed their file. Both worked. Frank was a steelworker, Susan a real estate agent. They had been married two years and three months. No children.

Frank had a full, neatly trimmed beard. He was almost six feet of solid muscle, broad-shouldered and with hands that were large and rough. Susan was a slim, beautiful woman with dark hair and large, expressive eyes, who, at first glance, seemed self-possessed and strikingly feminine.

Frank's entrance was commanding; Susan's was determined. They both had an issue and wanted to get right down to work. I motioned them each to a chair. Susan began.

"I'm grateful you could see us so soon. Frank and I have been having a difference of opinion and we need an arbitrator."

"Difference of opinion! Hell! You either abort the kid or it's over! That's all there is to it, lady." He was on the edge of his chair, one elbow on each knee and his finger was pointing at her. "You're not trapping me with this unwanted ____ !"

"Frank —" she began, but he jumped right in. "No! I won't have it! We either do it together or it's over. We had an agreement and you ____ well know it. *You've* done this. *You're* the one who said it wouldn't happen until we *both* wanted it. So get yourself taken care of. I don't want anything to do with this ____ problem!"

Susan turned to me imploringly. "You see, Dr. Woolf, how unreasonable he is—"

I looked beyond what she was saying and, with deliberation, interrupted, "Oh, I don't think he's being unreasonable. I think we ought to hear him out and—" with careful emphasis "—look deeper into the issues he's raising."

Susan caught right on and turned back to him and asked, "What are the *real* issues here, Frank?" Her transition was remarkable. She had changed from defensive to explorative without losing her composure. She was now listening totally. She shifted in her chair so that her body faced Frank and she was looking directly at him. Her eyes reflected total attention.

Frank softened immediately. He could see he no longer needed to shout and swear. He had finally got her attention. "Ahh . . . I don't know," he stammered and lowered his head. "I just don't think being trapped makes for good loving."

"But no one's trap— !"

"So, what you *really* want, Frank," I interrupted, looking at Susan but directing my voice to Frank, "is for your relationship to be two-sided. You want *both* of you to make the decisions."

I wanted to zero right in on Frank's positive intent. It's a sure way to end a fight.

Frank looked up. My steady gaze at Susan told her to listen and observe another way. I shifted back to Frank. "I don't want any kids right now," he said. "I can't handle it. It's just not in my cards right now. . . and NOBODY'S going to force me into it either!"

The "force" issue was very potent to him. He became immediately incensed when he thought he was being forced. I looked at Susan and said, "So, Frank, when was the first time you were forced by a woman to do something that trapped you?" I said it gently and quietly, but its impact was like a bombshell on him. His head came up and he said, "My God! I never thought about that! I, I was—" he looked at his wife.

"It's okay, Frank. I can handle it. I love you more than anything. More than the past." She waited.

"Well, in high school I had this girl. She wasn't really my friend but she really wanted to get laid and all the guys had had her, so I thought, why not? Well, she had some kind of racket. She worked me over real good. I was really sweating. In fact, I'm hot now. Is the room temperature up, or what?"

"No, Frank. The room temperature is the same, but you took a lot of heat from her, didn't you?" I asked.

"Well, she said she'd never done it with anyone else. I didn't believe it for a minute, but then she wanted more and she wouldn't look at anyone else. I just sort of got into it pretty heavy, and then one day she said she was expecting. I was only half there in the first place. I nearly croaked. By then, I knew she didn't have anyone else. She wanted to get married! I really sweated. I quit school and got a job. In a way, I wanted to anyway. She was all over me and I knew the harder she pushed, the less I liked it. I finally rebelled and wouldn't see her anymore. I just sort of went with the guys at work. I guess she found someone else because she never did have the baby. I wasted a year of my life! I was so stupid!"

"So you quit school and got trapped at the steel mill?"

"Well, it's not so bad."

"That's how you feel on the surface. Let's take a look at how that high school kid in you feels. Can you see yourself at sixteen?"

He settled back in his chair and relaxed. He said, "It's cool. I got my friends. I hate school. I got this English teacher that really sucks. She hates my guts, I know. My shop teacher is cool, though. He says I got natural ability with my hands. I figure I'll cut the crap and get a job and do what I want."

"Ask the sixteen-year-old how he feels about the girl."

"Jenny? Oh, she's cool. She's hot. She can't leave me alone. Whoowee! She's a real sport. But she gets her hooks into you and you're dead."

"Why is that?"

He had switched back to the voice of himself at sixteen.

"Cause she just wants a body. She wants a man to take care of her."

"So what's wrong with that?"

"Hey. Marriage is for love. I don't want to end up like my ol' man — " He stopped in mid-sentence. His head came up. "Oh, hell. My ol' man warned me, time and again. Never trust a woman who wants you to do it. Mom died when I was twelve, so I guess he never felt he got a fair shake. He never talked down about her, but he was bitter about women in general. Never did remarry."

"So he may have felt bitter that your mother died and left him to be both a mother and a father?"

"Yeah. He's a bit weird about it."

"In what way?"

"Well, he loved her. No doubt about that. He really loved her and he never remarried. But he didn't feel like he was good enough to be a parent without her. He sort of just let things happen with me. Like he lost his power."

"How does your sixteen-year-old feel about that?"

He bowed his head. "He hurts like hell. He's crying because his dad doesn't care. Nobody cares. All Jenny wants is for him to take care of her. She doesn't care about him."

"He feels trapped?"

"Yeah. If his ol' man can't be a father, how can *he* be one?"

We talked back and forth with the holodyne of his sixteen-year-old. We discovered how much he cared for his father and how much he missed his mother. We also discovered that he wanted, in part, to be taken care of, and could not resolve Jenny's need to be cared for with his *own* need. His sexual drive was fired by his need for love and connectedness. He was out of control. So he panicked, quit school, and ran.

As Frank became more familiar with this holodyne, he began to realize how much power this memory had upon his present attitude about his own fatherhood. He would keep exclaiming, "But I feel this way now!" and I would agree and point out how powerful his sixteen-year-old still was in his life. It dawned on him that what he really wanted was genuine intimacy with his wife. It also dawned on him that he was continually trying to control the relationship. He wanted control.

Finally, I suggested he imagine what it would be like to have a mature holodyne to help create genuine intimacy with Susan and bring him into control of his life. I suggested he imagine what such a guide would look like. He immediately imagined a being of light he called "Jesus." He was awestruck at the power with which the being drew near to him. Frank's countenance seemed to change.

"I feel His love," he whispered. "I feel His power. It's like Mom used to talk about. It's like Mom had with Him." He cried.

"Will he work with you?"

"Oh, yes. He will."

"Introduce him to your sixteen-year-old. Ask him to love that sixteen-year-old, to hold him in his arms, to accept him for all he has done and then get him to ask the sixteen-year-old if he is ready to take the next step on his journey to gain love and control."

We then had the holodyne of "Jesus" completely absorb the holodyne of the sixteen-year-old. We committed the new holodyne of "Jesus" to "be there" for Frank from that time on, to help Frank create genuine intimacy, to teach Frank all he needed to know about really loving, and to do all this based upon the principles of fairness, caring, and sharing. We created a place of peace and love in which "Jesus" could always be found, so that Frank could have him as an internal guide any time he wished. We then had Frank ask "Jesus" to help him resolve all the issues surrounding Jenny, his mother, and his father. Frank had a wonderful adventure with each person and was able to bring a new, mature, loving perspective into these relationships. He let go of his guilt and inadequacies, and even let go of his need to control.

Then I asked Frank, "Ask 'Jesus' about the baby."

Frank hesitated only a moment. He closed his eyes and said, "He says don't worry about it. He says I'll be a great father and he'll help me with the details." Susan came out of her chair and jumped on Frank's lap. "And so will I!" She kissed him.

"Will you come and be with me in the hospital?" she cooed.

"Hell, no!" he threw his arms wide, away from her. "I don't want any of the blood and all that stuff. But I'll compromise. I'll be there just before and just after!"

She was a little disappointed and she got up from his lap and sat down in her chair.

"Is there anything Susan can do for you?" I suggested. He thought for a moment and then looking a little sly, said, "Yeah, she can be there for me sexually."

"I'll compromise!" she challenged. "I'll be there just before and just after!"

We all laughed. "What does your guide say about all this?" I asked.

"Oh, hell," he chuckled. "Let's go for the whole ball of wax."

They left with their arms around each other.

Eight months later, I got one of those little "new baby"cards. In it was a picture of a beautiful baby girl and a note that said:

Thank you. We ended our adventure with compromise and have received the blessings of Heaven. May Jesus always be with you as he is with us.

Eternal love, Susan, Frank, and little Pamela

Tracking Summary

STEP ONE: ACCESS THE HOLODYNE

What is the problem? What feelings do you experience?

Do you have an intuitive guide who will help you access the holodyne(s) involved? Who will be your guide? Using your Intuitive Sensory Perception, re-experience the situation and feelings regarding the problem and have your guide locate the holodyne which is creating the problem.

What color is it? Does it have a shape? Can you "sense" it? Will it talk to you or communicate in some way?

STEP TWO: BEFRIEND THE HOLODYNE

Ask the holodyne what it has been doing for you. What does it want? Explore, in a loving way, its real intent. If the holodyne had what it really wanted, what would result?

STEP THREE: TRANSFORM THE HOLODYNE
INTO ITS MATURE IMAGE

Ask your guide to show you a mature holodyne which will represent what is really wanted. Will the mature holodyne help you? Is the immature one willing to find a better way to get what it wants?

Have the immature holodyne "grow up" into the mature one. Will the mature image commit to help solve the problem?

STEP FOUR: CREATE INTERNAL ORDER

Ask the mature holodyne if it is willing to teach you everything you need to know in order to solve the problem. Is it willing to dedicate everything it is to your fullest potential?

STEP FIVE: ESTABLISH A PRINCIPLED PERSPECTIVE

Will the mature holodyne always function on a principled basis? Will it be fair, care, and always share? Will it abide by the principles of equal rights and cooperation?

STEP SIX: UNIVERSALIZE

Will it sit at your round table, in your place of peace, and help you establish cooperation among all other holodynes and help you potentialize your life so you can act in accord with its new perspective?

Will it accept assignments and give whatever information is needed in order for you to completely overcome your old problem?

4

Potentializing

Life is abundant with enfolded potential. Even during the winter season, when growth is suspended in nature's icy grip, every seed awaits its springtime release. Every person, every circumstance contains hidden potential awaiting release. Like a seed frozen in the earth, this potential awaits its unfolding—its "potentializing."

Potentializing embraces all life processes, all growth experiences, all activity which releases and unfolds potential. Potentializing is a lifelong process: you grow, you experience life, and as you do, you unfold your potential. The seed within you, your "I," unfolds: your personality, your talents, and your knowledge awaken and develop according to your life mission. Like the caterpillar that evolves from an egg, spins its cocoon, and emerges as a butterfly, you evolve from one stage of your development to the next until you emerge as the highest expression of your Full Potential Self, your "I."

Think of the potential within a single apple seed, one small seed, to produce food for thousands yet unborn. Think of *your* potential as a human being. Who can measure it? Who can measure the potential of one person, one "I," let alone the potential enfolded within one relationship or one system? Who can measure the potential of one idea or one principle, let alone the potential enfolded within one *world* of people living together, working together, and learning together? And yet *all* these potentials, from the smallest to the greatest, are inseparably joined in our holodynamic universe: all draw upon its infinite power and possibilities and all unfold according to its Implicate Order.

How do you get what you really want—health, vitality, friends, love, success, peace, integrity? How do you break free from your limited thinking and realize your fullest potential? You learn to do, consciously and effortlessly, what it has always been your *nature* to do. *You align yourself with the holodynamic universe and you potentialize.* This is the deepest secret of success. This is why understanding the principles of holodynamics, holodynes, the holodynamic plane, and the role of the "I" is so important.

You cannot unfold your personal potential in its fullest sense without first unfolding the potential of every holodyne within your internal world—without *tracking*. And you cannot unfold your personal potential in your external world without unfolding *everyone else's* as well. Your potentializing affects, and is affected by, the whole of humanity. But you don't have to wait. By choosing to unfold, you change the physics of the whole field—you help unfold the holodynamic universe.

Potentializing and tracking are *conscious* applications of the same *natural* process: the unfolding of all life-forms according to the Implicate Order of growth. They differ largely in their focus. Tracking focuses *inward*, on the inner world, and potentializing focuses *outward*, on the outer world. The two processes are intimately intertwined. You potentialize by tracking primitive holodynes, and you track primitive holodynes by potentializing them. Every immature holodyne, no matter how powerful or how effective it is in blocking your progress, has a positive intent. It has a potential awaiting to be fulfilled. When you track that holodyne, you systematically potentialize it.

As your holodynes mature, your thinking and your behavior also mature. You realize who you really are and you find it easier to express yourself, to be yourself, to assert yourself, to declare what you feel, think, and want to do. You gain confidence. You become uniquely creative. Your deep, inner qualities and strengths—your love, intelligence, sensitivity, creativity, and vitality—begin to manifest themselves in your personal life. In short, you begin to *potentialize.*

You can clearly see the process of potentializing by phase-spacing and using the Mind Model. Phase-spacing allows you to step outside your perspective of the moment and see everything from a quantum perspective. From this perspective, your mind exists and operates within the holodynamic plane, a quantum field with both "wave" and "particle" dimensions. The "wave" of the plane is our whole, collective mind; the "particles" are the holodynes, among which your primary holodyne, the "I," orchestrates your life and responds to your conscious focus and your choices. Your conscious focus and your choices, no matter how subtle, impact the holodynamic plane because your thought-forms—your holodynes—have causal potency. They impact the Quantum Wave and tap into the Quantum Force according to the boundary conditions set by the collective "I." As you potentialize, you impact the whole field. Everyone around you is affected and all dimensions of the universe—past, present, and future—respond.

The Six Stages of Potentializing

When you align your conscious self with your Full Potential Self, you cannot fail. Your personal success is guaranteed. Then, as *you* potentialize, so also do your relationships, systems, principles, and even the universe itself. The six stages of potentializing follow, naturally and easily, the six stages of development—the Implicate Order of growth. You must align your conscious self with your Full Potential Self, your personal being with your relationships, your relationships with your systems, your systems with your principles, and your principles with the holodynamic

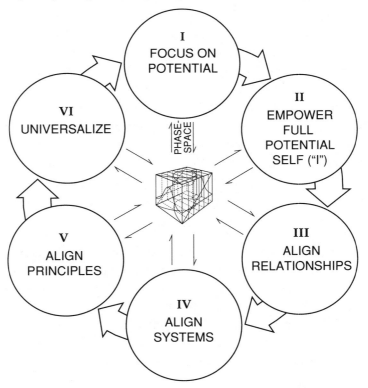

FIGURE 11: THE SIX STAGES OF POTENTIALIZING

universe. When you do this, growth takes place naturally, effortlessly, without resistance.

The stages of potentializing apply across the full developmental spectrum of the Mind Model: they apply to your holodynes, to your personal life, to your relationships, to your systems, and to your principles. Following these six stages guarantees that you will realize your *fullest* potential—both personally and universally: that you will maximize your impact upon the holodynamic universe.

STAGE ONE: FOCUS ON POTENTIAL

Everything has enfolded within it a hidden potential—every person, every situation. To unfold it you must first learn to focus on what that specific potential is. Your focus depends upon your sensitivity—your mind can only receive information to which you are sensitized.

Attune Your Senses

Information coming in through your senses is processed by your central nervous system and stored in your brain as holodynes. Information *outside* your comfort zone is usually negated or discounted by the holodynes of your family and cultural beliefs and, in effect, *blocked* from your consciousness. You cannot "see" another person's point of view, you cannot "feel" close or sympathetic to another because you are locked into an old belief system which makes you insensitive to that person, or which continually tries to get him or her to change. You attune your senses by potentializing—tracking to their fullest potential—the primitive comfort zone holodynes which are *blocking* your senses, and keeping you "in the dark."

Use Your I.S.P

You focus on a specific potential with your intuitive senses. In other words, you use your I.S.P. Your I.S.P. is the key to finding not only your problem holodynes but their hidden potential. To engage your I.S.P., simply begin from your place of peace and ask your Full Potential Self, or some other mature intuitive guide, to help you. Then focus on the potential and "sense" it with *all* your senses.

If you are relating to your spouse, a co-worker, your boss, an employee, or anyone else, always keep your focus upon the positive intent of that person's fullest potential and reflect back to him or her the intent you sense. Ask yourself:

"What is it this person really wants?"

"What is the best possible solution to this problem?"

"What can be done in this circumstance so that everyone wins?"

Phase-space

Focusing means becoming attuned to the real potential enfolded within a person, relationship, situation, system, or principle and then *phase-spacing* that potential to see it clearly. With phase-spacing, you can act rather than react, solve problems rather than become part of them, and bring into your life almost any dynamic you choose.

Suppose someone in your family is depressed—your uncle, for instance. He isolates himself, doesn't want company, and withdraws into his own private world of blackness. What can you do? You can go to your place of peace, call up his Full Potential Self and ask for help in finding the underlying order within his depression. Somewhere along the line, your uncle stepped off the path of his potentializing: somewhere he decided not to take responsibility for his potential and, because his life energy then turned inward and downdrafted, he became depressed.

Using your I.S.P., you can gain a real sense of the hidden

potential within your uncle's depression. Now you can go to him with confidence; you can help him find the potential of his depression. But remember, you can only help. It is *his* path and *his* potential. You can accept where your uncle is, you can walk lovingly with him until he finds his way back to his path, but you cannot lift him out of his depression. Only *he* can do that. His depression—*every* depression—is an unfulfilled potential waiting to unfold.

Once you have identified the specific potential, you must *keep* the potential clearly in focus—no matter how much opposition or turbulence, no matter what negative dynamics, you encounter on the holodynamic plane. Keeping a clear focus on the potential is an act of intelligent love: by engaging both the rational and intuitive processes of your mind, you activate the quantum field with regard to that potential. You can do this easily if you know how to phase-space, using the Mind Model. Thus, once you have focused on the potential, you identify its rational and intuitive aspects and you place it at its stage of development within the holodynamic plane. This completes Stage One: you have identified the potential, located it, and brought it clearly into focus.

STAGE TWO: EMPOWER YOUR FULL POTENTIAL SELF

Once you are sensitized to the specific potential and have it clearly in focus, you can check out the boundary conditions by calling forth your Full Potential Self. You can ask such questions as:

"Is this potential possible in this world?"

"What are my specific responsibilities or tasks regarding this potential?"

"What specific responsibilities do others have regarding this potential?"

Almost any new potential will create resistance in the holodynamic plane. Using your Full Potential Self as your guide, locate those resistances *within yourself* which are telling you this potential cannot be reached. Ask yourself:

"Which of my holodynes are blocking the fulfillment of this potential?"

"What problems, challenges, concerns, or limitations exist which might limit or curtail fulfillment of this potential?"

With your Full Potential Self, you can address each block and determine the best possible way to handle it. In this way, your conscious self and your "I" are always fully aligned. You work on fulfilling the potential together. You team up. Your teamwork creates the synergy needed to accomplish any project. You enter into it knowing yourself, completely aware of your personal mission, feeling confident and free to be creative in unfolding the potential which you have in focus. Your Full Potential Self, from the deepest order of organization, makes sure the potential is aligned with all other aspects of the quantum field. And your task becomes easy and natural.

As you track and potentialize all the holodynes resisting the potential in focus, their combined, new energy amplifies that potential's causal potency. The holodynes on the holodynamic plane within your mind come into harmony with you, with your "I," and with your chosen potential. This triggers the Quantum Wave. Phase-spacing helps keep track of all the holodynes and identifies any saboteurs to the process so they can be tracked. This clears your channels and leaves you free to commit with your whole being.

STAGE THREE: ALIGN RELATIONSHIPS

Once you are free to commit, you can establish the necessary rapport with others, build a foundation of respect and friendship, evaluate skills, and develop whatever intimacy is necessary to accomplish the unfolding of the potential. Your commitment is clear and, in that clarity, you can find others who are aligned, or can be aligned, with that same potential. You ask yourself:

"Who do I need to work with regarding this potential?"

"Who is aligned with the potential and who will team up with me to fulfill it?"

"What channels of communication need to be established?"

"What tasks and duties need to be assigned?"

"What needs to be learned and what skills developed?"

Use your I.S.P. to access the holodynes of your fellow workers. Your clarity and intuitive skills will maximize your ability to align your and their Full Potential Selves with the potential in focus. You may need to teach the others how to track their own resistances and how to potentialize. When you are all aligned, your rapport, cooperation, and teamwork reach a peak and every relationship within the system becomes empowered with regard to the specific potential. The whole system becomes activated.

Phase-spacing relationships allows you to immediately identify all relationship dynamics. You can see the level of commitment of each person. You can identify downdraft games and recommend appropriate action so as to allow everyone the fullest opportunity to participate. As each relationship aligns itself with the potential, each becomes part of a living network, part of the living dynamic of the system, and the whole system comes to life.

STAGE FOUR: ALIGN SYSTEMS

Now you and everyone else within the system are free to *act*. All of you can now plan your work and work your plan. Questions to ask your team members:

"What are the essential ingredients that make up our plan?"

"What equipment, supplies, facilities, and physical resources do we need to accomplish it?"

"What skills, talents, and other personal resources do we need?"

"What kind of production schedule seems feasible?"

"What job descriptions, duties, and contracts do we need to draw up so that we can all work in harmony?"

"What lines of authority and specific stewardships must we have in place for the whole system to function efficiently and effectively? How can we maintain maximum productivity and harmony?"

As the system comes alive, all facets are attended to by their respective stewards, adjustments are made and fine-tuned, and all parts of the system are aligned so that the potential of the system is fulfilled. Because the value of the objective, the potential in focus, is known throughout the system, because each Full Potential Self is aligned with reaching that potential, each person becomes pro-active, self-activating, totally committed to the action at hand. And as each person fulfills his or her part, the system fulfills *its* part, rewarding everyone accordingly. It may be winning a hard-fought ball game, staging a dramatic hit, breaking into a new market, or creating a new product. Being part of a successful team gives you a special feeling of accomplishment.

Use phase-spacing to get a clear view of the whole system's dynamics. Keep your eyes open for any downdraft processes and see that appropriate alignments are made. The system is now ready to become self-perpetuating—to establish and align its own set of guiding principles.

STAGE FIVE: ALIGN PRINCIPLES

Most people, projects, and systems fail because they lack guiding principles and, lacking them, they drift helplessly onto the rocks, like ships without rudders. To survive and to prosper, your system must have a set of universally aligned guiding principles in which you all believe. You cannot, for example, expect to have plenty of money unless you understand that money comes as a natural reward for value rendered. The greater the value you render, the greater your reward. If you expect money without developing the gifts you have which are of value in life, you are out of alignment. You believe in getting something for nothing. The principle of something for nothing is a downdraft principle and will cause you endless backwashes and whirlpools in the streams of life's abundance.

Abundance, health, peace of mind, intimacy, success of any kind are all guided by deep, underlying principles. Principles are like irrigation ditches which guide the flow of water into a farmer's fields. Only when the ditches are properly aligned will the water reach its intended destination and the crops prosper.

One major company has developed just such a set of aligned guiding principles, as aptly illustrated in the following company story. A secretary was waiting in line at the copy machine when the man in front of her made two copies and then walked away. She called him back, explaining that she did not want her department charged for *his* copies. He apologized and signed for the copies. When she looked at his signature, she saw it was the company president's. She started to apologize but he would have none of it. He said, "You were perfectly right. I was in the wrong." And next week she was promoted. In that company, the principles of equal rights, employee integrity, and honest, open confrontation cross all lines of authority. They set a high standard of responsibility for everyone, but they pay off, and the company performance records show it.

Any system's potential can be aligned with the principles of fair treatment, of sharing, of genuine caring for all system participants. Such principles reflect the resonating energy field which gives life to the system. Whether a system lives up to its principles can readily be seen in how well everything is working. Questions to ask:

"How efficient is the system?"
"Is it meeting the needs of all its participants?"
"Is the potential being reached according to plan?"
"What could be done better?"
"Do we need to adjust our plan?"

Use phase-spacing to locate any breakdowns in the integrity of the system, any violations of the system's guiding principles, either within itself or between it and other systems, and to find the best remedies. See that all holodynes, relationships, and parts of the system are aligned within the guidelines of the system's principles. You are now ready for the final stage of potentializing, which assures that the potential can in fact be realized.

STAGE SIX: UNIVERSALIZE

Once all of its internal dimensions are aligned, the whole system can *extend* itself; it can project its own causal potency throughout the holodynamic universe. Its impact can be felt upon the larger community and all decisions with regard to reaching the specific potential must be evaluated in light of the larger community. Questions to ask:

"Does this potential represent the highest possible use of resources for the larger community?"

"Are we using these resources in the best possible way for all concerned?

"Are there any other larger dynamics we need to consider?"

When you universalize, you lose nothing and gain everything. You receive open input from every level of a system and of the larger community in which the system exists. You use your rational mind to analyze any particular dynamic and your intuitive mind, your I.S.P., to phase-space the whole dynamic. You sense the potential of every holodyne, person, relationship, system, or principle at every stage of development in the whole dynamic. You become attuned and empowered, at one with the holodynamic universe.

How do you achieve this in your life? You begin at your place of peace with your Full Potential Self. You allow your "I" to unfold, using your I.S.P. and phase-spacing. You learn to focus, to align your Full Potential with your focus, to align relationships, systems, and principles holodynamically. You *potentialize*

In the chapters that follow, I will show you how simple, how powerful, and how easy all of this is. I will do this by presenting certain themes and by showing how the inward process of tracking and the outward process of potentializing work together to help you unfold your fullest potential and realize the "I" that you are.

5

Physical Well-being

On the street, as a therapist, I saw a daily stream of people who were sick, hungry, destitute, and desperate. It didn't take long to discover what had made the difference in their lives—why they were sick, while others were healthy, why they starved, while others prospered. What had mattered more than anything else was what they *thought*, what they *believed*, about themselves, about their possibilities, about life itself.

Someone once said that if you divided up all the money in the world so that everyone had an equal share, within five years it would be back in the hands of the people who had it before. What causes one mind to create deprivation and another abundance? How does this happen? What are the rules of the game being played? Can we look at the orders within the orders and access the holodynes responsible for illness? Poverty? Chaos?

The "I" within you, your Full Potential Self, is the most powerful agent for change in your physical world. When you align your "I" with your conscious choice or focus, you stimulate the

Quantum Wave to produce an echo effect in your Full Potential Self and this, when the boundary conditions are met, collapses the wave into the physical reality you have chosen. So human intelligence is at the very center of physical reality. In order to understand this physical world and all its ills, you need to understand the "particle" aspect of reality in a rational way, the "wave" aspect in an intuitive way, and the holodynamic plane, where the "I" functions, in a way which brings both the "particle" (rational) and the "wave" (intuitive) together into one, whole dynamic.

Social ills have their origin in the quantum dimension, where all things are made of energy, where physical shape is secondary to hidden, powerful fields of influence, and where human creativity is central to everything. Human creativity is the catalyst which precipitates the collapse of the Quantum Wave into physical reality, the instrument through which the boundary conditions of reality are set, and the deepest reflection of the Implicate Order of reality. What part, then, does human creativity play in abundance, health, vitality, and other conditions of physical well-being?

You create your physical condition.

Does that mean when a little baby gets sick and dies, the baby "knew" what it was doing? Does that mean when some big government declares war and countless thousands are forced to suffer, starve, and die, those thousands "asked for it"? Does it mean that every person at every moment and in every way is getting exactly what his or her "I" has chosen? Yes. That's just what it means. There are no victims. Poverty, suffering, disease, and death are the result of choices made by the "I," which knows everything throughout the quantum field underlying all reality—past, present, and future.

"Fault" is never the issue. Finding fault is simply Stage Four downdraft thinking, which is judgmental as a part of its "natural" disposition. When you are a judger, your reward is a feeling of righteous indignation toward anyone who does not believe as you believe, or act as you think is right; you have little, if any, responsibility for those you judge, and even less choice about them. The resulting turbulence in your resonating energy field may not seem to have much effect upon you—you may feel "comfortable"—but you are likely to suffer in other ways. You may, for example, have trouble "connecting" with others or having fun, you may be lonelier than other people or find yourself preoccupied with other people's business. For when you judge, you find it easier to address everyone *else's* problems, and so you become filled with dis-ease. You are seldom, if ever, really content. You can't see that suffering is just a "built-in" consequence of judging.

To the mature mind, "lessons" and "choice" are always the issue. From a Stage Six updraft perspective, you can understand what your suffering is all about. The "I" within you knows the past, present, and future and has designed your life experiences in response to your choices. Physical health, wealth, and life energy are the result of choices you make deep within the holodynamic realms of your mind.

Using Intuitive Sensory Perception, you can reach these deeper dimensions of your mind and choose to change the whole dynamic—to impact the holodynamic plane. You can overcome poverty, restore health, and increase vitality. But first, you must understand how the real world works, how your mind can impact this world, and how each dimension of your mind sets the boundary conditions by which you can collapse the Quantum Wave and create reality.

The Mind Model gives your rational mind a map by which to integrate the "particle" functions of your holodynamic plane at every stage of development. Your intuitive mind, using I.S.P., creatively manages the deeper "wave" levels of your holodynamic plane, where real change takes place. Through tracking, you can induce any immature holodyne which may be causing disease, addiction, suffering, poverty, or the like, to complete its own growth, to release its fullest potential into your holodynamic plane, and, by changing the resonating frequencies of the plane, to generate new growth within your life. This will happen when the time is right, and not necessarily when you *want* it to happen. Remember, holodynes work on a different order of time from ours.

Quantum thinking, I.S.P., the Mind Model, tracking, and potentializing can work for you, now, in the world of reality, your physical world, so that you, *by your own choice* can establish a greater degree of physical well-being.

Your Body

From a holodynamic perspective, your body is organized within a field of resonating frequencies which create both its "wave" and "particle" functions. The source of life energy is the Quantum Force. It is primarily your "I," acting like a slide in a slide projector, which determines the particular shape and size of your body. Your "I" draws the Quantum Force from the unmanifest plane and collapses its "wave" function to form the pattern of your body. This takes time—at least nine months in the womb and years of growth to reach your prime—and can be influenced by environmental conditions, but from a quantum perspective, it originates deep within the quantum field—with your Full Potential Self.

As your Full Potential Self, your "I," draws the Quantum Force through itself, it sets up a specific field of influence, which resonates according to certain frequencies. If we follow Jean Piaget's and Rupert Sheldrake's line of reasoning, these frequencies encode instructions into the subatomic world so that atoms form in the field. Once atoms form in sufficient numbers, their resonating frequencies set up a biochemical field of influence so that more and more complex molecules can form, and thus eventually DNA and RNA. These genetic molecules, in turn receiving and transmitting encoded frequencies from the "I," are able to form living cells. And so the process continues, from cells to tissues, to organs, and finally to your body as a whole.

Diagrammatically, this can be seen below.

Scientists can already detect, indirectly, some of the resonating frequencies which emanate through the various levels of the body. All cells of the body, for example, can be viewed as receiving and transmitting resonating frequencies. The same is true of tissue and organs. Even though we lack the necessary technology to observe these frequencies directly, at their point of origin, we can deduce that they come from some central source, which I call the Full Potential Self, the "I." Since the frequencies can be detected within the body at the levels of cells, tissues, and organs, I am assuming they also function at the molecular, atomic, and subatomic levels, as well. Based upon the evidence coming in, this

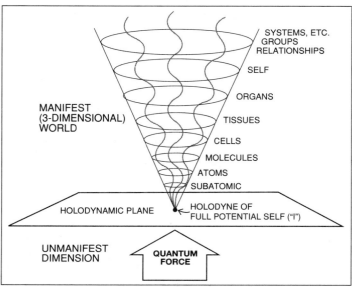

FIGURE 12: CHANNELING THE QUANTUM FORCE

is a safe assumption. So that at all levels of organization, from the subatomic to the organismic, your body receives and transmits encoded messages.

Each stage of your development is governed by a deeper order, the Implicate Order, embedded into the fabric of space and time. And each stage unfolds within a field of influence which, I contend, is programmed by your Full Potential Self, your "I." The encoded resonating frequencies of this field set up by your fullest potential orchestrate the whole dynamic of your development and extend outward in every direction. So you are part of a much larger field, you are part of the holodynamic universe.

Your personal resonating frequencies do not end with your body. They continue throughout the whole quantum field of reality. This is why Cleve Backster's experiments detect an instantaneous connection between a person and a single cell from that person even though they are miles apart. This is also why people who are emotionally bonded can sense what is happening to one another, even though they may be in different parts of the world. Time and distance do not seem to affect the field because it is *pre*-atomic. It is part of the quantum world.

Disease

Disease first enters your body at the level of holodynes. Every disease is caused by a series of holodynes, which impact the quantum field with the particular resonating frequencies of that disease. Your "I" allows these disease holodynes to affect your personal field: disease is always the result of some lesson you need to learn, the consequence of some choice you have made at some point on the holodymamic plane of your mind. The disease holodynes always know what the lesson is and what choice created the disease's power in your field. Every disease holodyne is aligned with the consequences of choices you have already made.

Medical experts almost universally agree that over eighty percent of disease can be traced to some psychological cause. What, then, is the mechanism by which the mind creates disease? It has to do with the orders within orders of the mind—with holodynes. And how do we maximize our ability to heal? We use the Mind Model and phase-space the holodynamics of physical well-being.

So, do you want to be healthy? Do you want to be rid of some disease? Let's go through what you can *do* about disease, using the Mind Model, phase-spacing, I.S.P., tracking, and potentializing.

First, decide whether you want to look at your disease from a rational or from an intuitive perspective. If you begin with the rational, you could study everything you can get your hands on about the disease—medical causes, historical facts, possible cures. If you continue on the rational side, you could investigate your personal history to find out what was happening at the time you contracted the disease—what "burning issue" might be associated with it. You could ask yourself whether somehow your mind subconsciously connected the two. If your "burning issue" is still unresolved, your disease may be an extension of it.

You could look further, into your family histories, your associations, and your work environment, into your religious, political, or economic issues, to see whether there is any possible connection between what is happening in the whole "field" of your life and this disease. When you complete your analysis, you could simply try out the cures. If they work, you have what you want. If they don't, you may want to go into the orders beneath the orders. To do this, you must shift to your *intuitive* mind.

Once you decide to go "in-tu-it," begin from your place of peace. "Become" peace and call forth your Full Potential Self or some other intuitive guide. Then, using your I.S.P., enter into your body and sense the presence of the disease. Envision it within your body. Feel the disease, experience it, unveil its resonating energy field. This means really "being there" for the disease and learning everything about it from your body. Next, pull the disease out and examine it as a living entity. Befriend it, find its positive intent, what it would be like at its fullest potential, and then create a new, mature holodyne to represent that potential. Have the mature holodyne absorb, or transform, the immature one.

In other words, *track* the disease. Complete the six steps of tracking by having the new holodyne commit to be there for you, to fulfill the intent of the old, immature holodyne, and to sit at your round table in your place of peace, so that you two can always communicate. Ask your new holodyne to teach you everything you need to know about this disease. Follow through. Learn what you need to know. Give yourself space for healing. Respond to your intuitive guides. Eat what they tell you to eat, go to the doctors they tell you to go to. Take charge of your healing. In the meantime, let your rational mind continue to correlate your analysis with all medical knowledge and every other fact known about your disease. In this way, through the balanced synergy of *both* sides of your brain, you can maximize your healing.

So, how is it that some people always seem to be healthy and other always sick? How is it that some people always seem to have lots of money or vitality and others never enough? Why do some people have a weight problem and others not? Where is the key to all of this? The key to health, weight, abundance, vitality, and physical well-being lies in how your mind actually stores its information, organizes it, and then programs your life—on how you deal with reality.

Mind-Body Unity

If you want health, if you want a body which has lots of energy, a body free from disease or addiction, you need to understand that your body is an extension of your mind. We know that the central nervous system is involved in the storage of holodynes throughout the body. For instance, you may have experienced the release of all kinds of emotions when a massage therapist worked on a certain spot of your body, like your knee or shoulder. But your body is more than just a storage place—it is a multi-dimensional extension of your mind.

Your body and mind are one.

The idea that there is a separation between your body and your mind is an illusion, perpetuated by holodynes which believe in separation. Separation thinking is downdraft thinking: it keeps you feeling discontent, isolated, "outside" yourself, always looking for a way to separate yourself from your body in order to find your spiritual life. You can immediately overcome this sense of separation by dealing with the holodynes at Stage One of the Implicate Order, by entering into the realm of holodynes that are in charge of your body, that are in charge of your weight, your size, your wellness or illness, and even how much energy you have. These holodynes are connected to *all* of your physical circumstances, they handle all dimensions of your physical world.

If you feel separated from your body or from any part of your physical world, then you most likely have holodynes that believe in separation and want you to experience the lessons to be learned from being separate and lonely and miserable and isolated. If you've learned enough from those downdraft feelings and would like to updraft your physical body and become healthy again, then you need to access the holodynes that believe in separation.

Separation holodynes usually intend that you should find unity again. The irony is that unless you choose to mature these holodynes, to track them and potentialize them, unless you *stop* trying to escape from your body, you may never discover your unity. You are an "I" having a physical experience, not a physical being trying to find its way back to its "I." When your physical body becomes a true temple, when it becomes part of the "every-where-everywhen" dimension, you realize that "here" is all you've really got, that "here" is the *only* place there is. There is no other place to go because everything comes from your Full Potential Self. All other lives, all parallel worlds, all holodynes from other realms are centered on your full potential *here*, in your body, now, at this time. So the unity you seek cannot be found by seeking other worlds. It can only be found by potentializing, by aligning your conscious self with your Full Potential Self and letting your intuitive guides help you unfold the unity you already are.

Every disease you have had, or now have, is there to teach you lessons. Come to be "at one" with your dis-ease, track it, and potentialize it. Learn all the lessons that it has to teach you. Then let it fulfill its fullest, highest potential. If you have a disease that usually causes death, this does *not* mean that you have to die. Dancing on the razor's edge of death is meant to teach you deep appreciation of the tremendous value of life itself. Once you have learned this lesson, then you can mature the holodyne and let it teach you the value of life in a *mature* way, rather than by having you dance on the razor's edge, always wondering whether you're going to live or die.

When you focus on life instead of death, when you choose to put your energy into life itself, instead of into dis-ease, your choice sets up a healing field of resonating frequencies to which all things respond. Once you have truly matured the holodynes of your disease, then nothing can stop your healing. You have learned the lessons of your disease and you can move on to other lessons, perhaps in some other area of your life.

But you will never learn the lessons of life by escaping your body, which is your vehicle of life in this dimension. Therefore, stop trying to find divinity by dying. You are holodynamic—your body and divinity are inseparably connected. They're one in the same. In this physical world, your spiritual fulfillment comes through your body. Your intellectual fulfillment comes through your body. Your rational and intuitive selves both seek manifestation through your body. It's a matter of balance, of choice, of "becoming" part of the field. At the center of it all is you, your "I." All that you manifest here in this physical world is, in reality, the unfolding of your "I."

6

Personal Well-being:
On Overcoming Evil

What is Evil?

The only thing stopping us from living life to its fullest are our illusions. Foremost among these is what we call "evil." Evil is a *projection* of our downdraft holodynes, our negative illusions. As we overcome these illusions, we overcome evil—and reclaim our birthright of abundant life.

Evil is live in reverse.

Evil, not death, is the opposite of life. When we live in reverse, we invest in death rather than life, in the power of nothingness rather than the power of everything. Evil draws its power from the primitive holodynes of downdraft thinking at every stage of our mental development. The expressions we give to evil differ according to our rational or intuitive processes, according to the stage of development we choose to operate from, and according to the particular holodyne or holodynes we choose to express it with. So our experience with evil can have great variety and can influence any circumstance in which we find ourselves.

Now if we take evil through the six stages of development, we get a very different view of evil at each stage.

Evil to a Stage One mind does not even exist. At this stage, you may feel evil as fear in the form of repressed nightmares or unseen terrors, but because your mind cannot yet conceive of anything that is not physical, it does not recognize evil.

As your mind shifts to Stage Two, your personal ego takes over. Evil appears as anyone or anything that does not agree with your own private image or interpretation of things. Here evil represents all those forces which you feel are against you, which are trying to overcome you. You may experience evil as "the Devil" or "demons" who are trying to "possess" you, or it might be pressure, bad luck, or your own negative thoughts which drag you down and defeat you.

At Stage Three, you see evil in anyone who does not agree with your *family* or your small group—and in anything that doesn't match your *family's* images of what is right.

At Stage Four, you and the whole system to which you belong come to believe that *your* way—your church's, your company's, your country's—is the *only true* way, and that "anyone who does not think like us or join with us—anything that does not agree with us—is *against* us" and therefore evil. By doing so, you cut yourself off from everyone *outside* your system—from the truth that others have.

At Stage Five, evil becomes anything that *limits* you. Your mind makes a quantum leap. You discover, "*I* am my *own* evil"— that *you* made up all the "only" games, that *you* created the evil in your life, and that evil can only exist if *you*, the individual, allow it to. You learn that, by *owning* your evil, you can also *disempower* it, you can locate its holodynes within your mind, mature them up, and release the good which is their fullest potential: you can *track* and *potentialize* your evil.

At Stage Six, evil occurs when you choose *not* to universalize, when you place yourself *above* everyone else. You feel euphoric, enlightened, superior to others, and so, instead of extending yourself to them and living your life to its fullest *with* them, you detach. The more power you have, the more remote you become. You see evil in all who question you, in all who do not believe in you *absolutely*. Obsessed with your sense of righteous superiority, you demand subservience. You "speak for God," and, finally, you become an absolute tyrant. You *downdraft* your universal potential into universal evil.

To have any impact upon evil, you must understand that the life force which underlies it is *always* positive. You cannot "cast out your demons": they will only withdraw into your subconscious—to return in other forms. But you *can* transform them into their positive intent, you can *potentialize* the life force you have invested in them and release its good. To do this, to overcome your evil, you must first *love* your evil as a holodyne, a part of your mind that is trying to teach you a lesson. Then and only then can you find out what your lesson is and *learn* it. Then and only then can you take your "evil" holodyne through the stages of its growth and transform its power over you into good.

Because evil is a projection of our negative illusions, we can overcome it with truth and with love. Only when we "resist," when we feed it with our primitive fears, can evil persist and grow. Evil has no power over a mind which is mature. The games that primitive or "evil" holodynes play can no longer be played in a mind which has consciously chosen to move beyond them.

Living Beyond Evil

The power of evil arises from our *blocked* potential for good. Indeed, it is our experience with evil which teaches us most about our potential for good. By experiencing and understanding evil, we become response-able for our thoughts, our feelings, and our actions, we learn to move *beyond* our limitations, *beyond* our evil, and to empower the good we are and always have been—to unfold our "I."

To live beyond evil, you must make a passage. You must walk a path which leads into and *through* evil before you can reach the fullness of life beyond evil. This passage comes naturally when you choose to let your mind evolve—to own your evil and to go beyond it, beyond your own limitations.

The part of your mind which normally sees evil in others or within yourself cannot move beyond its own judgments, cannot move beyond the evil it sees. If you would live beyond evil, you must access the part of your mind which *knows* the evil within you, within everyone, which *knows* the positive intent behind every evil thought, feeling, and action, and which *knows* how to mature the evil up to its full potential for life. You must learn to trust your "I," your Full Potential Self.

Your "I" understands that the meaning of life lies *beyond* good and evil, that good and evil are only *part* of the masterpiece of life. Your "I" can see that neither good nor evil has any power but the power *you* give it through your holodynes. *You* hold the keys to the power of good and evil within your mind—in the holodynes you have created, received, or inherited.

When you *choose* to develop—to befriend and help your primitive holodynes reach their full life potential—you *transform* your evil into good. In your mature mind, evil becomes a positive life force: you change the resonating frequencies around you, you change the physics of the universe, from evil to life.

To pass beyond evil you must confront the "games" your mind has been playing which perpetuate evil. When you use your I.S.P. and phase-space these games, you can see that, for the most part, they stem from childhood fears orchestrated by your own self-limiting holodynes. Some evil may be inherent in your family and cultural beliefs, and some may be woven into the fabric of your most sacred beliefs. Some evil may come from parallel worlds, and some you may attribute to demons. Whatever its source, overcoming evil requires that you have a mature mind—and that you make a passage.

BEYOND CHILDHOOD FEARS

Your passage begins when you first confront evil. Your first evil is usually seeded in early childhood, when the world is new and your innocent mind attempts to make sense out of it, out of other people's behavior. You experience fear and terror, and from these potent feelings, your mind creates images. Beneath every bed and behind every door lurks a bogeyman or monster who is "out to get" you. These images become empowered holodynes. Thus are born your insecurities, and the "demons" of your family tree.

Your mind has begun its experience with evil, with the powers of darkness and entropy which lead to death. By making the passage from innocence to darkness, your mind comes to know its own dark side and, by contrast, to glimpse the light—to sense its potential powers for "good." It begins to polarize, to think in terms of "good and evil."

BEYOND SELF-LIMITATIONS

As your mind develops, your primitive holodynes express themselves as "evil forces" within you. Each time you attempt to accomplish some "good," these holodynes create a downdraft, a negative drain on your life energy. The seed that was planted at the very start germinates, puts down roots, and grows within you. This is why you are so comfortable with your limitations. This is why your evil endures. It becomes familiar. When not confronted and matured up, your old, comfortable, downdraft holodynes reflect themselves both rationally and intuitively in all your activities. They are the source of all your limitations and therefore of your evil.

How, then, do you overcome your "implanted" evil? You overcome it first by identifying the holodynes which are at its core, the hidden primitive programs of your mind, and then by accessing these holodynes through your I.S.P. They will actually reveal themselves if your mind is attuned to their existence and receptive to their messages. But if you are hostile or judgmental toward them, if they sense you want to cast them out or destroy them, they will hide. They will play great deceptive games and project themselves onto others around you. Thus you come to see your *own* evil as the evil in *others* and your own good as being associated with them.

Notice how much easier it is to see the faults of others than it is to see faults of your own, how easy it is to point out others' faults and how hard it is to hear others point out your own. All your limitations, all your personal "good and evil" games, will persist in your mind and in the minds of your children and of your children's children until someone, somewhere learns to feel compassion for the holodynes perpetuating the games, and *potentializes* them.

If you will create an atmosphere of unconditional love for *every* holodyne, no matter how "evil" it may appear, you will be able to understand, mature up, and thus overcome your evil. You must be willing, however, to face *all* your defenses, *all* your games of repression, projection, and avoidance, and see them as part of your own evil. Such games "help" you in the worst sort of way.

They keep you irresponsible, in "a-void" mode, so you don't have to deal with your problems, your responsibilities, your life. They make you feel good, even "righteous," without having to act on *anything*. Your evil just glides along unattended. Now, you can appreciate these games for what they intend. You can even love your games. Every part of your mind will respond to loving intelligence, even those parts you first judge as totally evil or primitive.

You will *naturally* feel unconditional love for your "evil" holodynes—once you get to know them and find that they are positively motivated. When you phase-space the holodynes which are perpetuating your evil, you will find that each holodyne was formed as a part of your childhood solutions to problems which your childhood mind created. All of your fears, all of your defenses mean to keep you safe. They don't, of course, but bless them for trying so hard!

All your evil acts and all the evil acts you will ever witness have, if you look deeply enough, a positive intent. The positive intent of your "evil" holodynes is to get rid of conflict, to be unified again. But they are limited by their primitive outlook. They want to return to the unity of the field by destroying all "wave" and "particle" existence. Their intent is good but their process is primitive.

As a teacher and a therapist, I have encountered "demons" or "entities" many times. And every time these primitive holodynes have responded to an atmosphere of unconditional love, have revealed their positive intent, and have allowed themselves to be matured to the fullest potential of that intent. Whether you believe in the devil or in demons or not, whether you believe that such entities exist in your mind or on some astral plane, is irrelevant. The fact is that no evil can have power over you in *this* plane except through the primitive, downdraft holodynes of your mind. Once you learn to access these holodynes, accept their positive intent, and *potentialize* them, you can transform their "evil" into good, their downdraft into updraft, and release the powerful life force within them. The choice is yours.

The Function of Evil

When we are locked into our own primitive perspectives, we cannot see our own solutions. We drop into our own mental "black hole." Our "evil" holodynes would have us accept their primitive reasoning as our own. For example, they would have us believe that only by getting rid of *all* causes of disunity can unity be truly found. Since wherever there is life there is disunity, life must be a *cause* of disunity and therefore, they reason, the only way to find unity is to *eradicate* life. Thus evil, in its quest for unity, expresses itself as the forces of darkness and death, and seeks the end of all life. Now this may seem preposterous but millions of people secretly, subconsciously believe this is as it *should* be. Part of *your* mind may believe it. Such beliefs help create the downdraft in your mind which acts against all the good that you would do.

The function of evil is to help your mind *polarize,* so that you can experience life more fully. From the antithesis of evil your mind is challenged to create the conscious thesis of good. You must consciously choose to honor the variety of life, choose to defend the freedom life gives you, before you can value your own life force and unfold the "I" within you. And it takes the light of good and the darkness of evil to complete the masterpiece of life.

From your very first stage of development, your mind polarizes: pleasure with pain, health with sickness, abundance with deprivation, life with death. Your physical well-being arises from your natural resistance to—from your *overcoming*—a tide of germs, stress, misery, and deprivation.

I had a patient who loved to stick pins and nails into himself. As we explored the holodynes involved, we discovered he had been criticizing himself, brutally and constantly, for as long as he could remember. Further exploration showed that his mother and father had criticized *everything* he did: he could never be good enough, never do anything right, he couldn't support a family and would "never get married, never amount to anything." They had all but driven pins and nails into him themselves.

If you are like most of us, you are continually talking down to yourself, continually trying to convince yourself you are less than you are. In this way your polarizations develop real power in your life. We come to earth with perfect intelligence and love, bonded together for eternity, but we must do everything we can to learn to appreciate this truth. So we talk down to ourselves, we criticize ourselves, and we continually try to convince ourselves we are *not* intelligent, *not* loving, and *not* bonded eternally. We fight each other, we blame each other, we withdraw from each other, and we avoid the truth about our love.

Notice how you perpetuate these "evil" games in your family patterns and in the games you play at church and at work. I had another patient, a woman, who worked as a police dispatcher and who was so demoralized by the demeaning comments, the dirty jokes, and the degrading atmosphere of her workplace that she wanted to commit suicide. Her life had become a mess. She had been listening to this stream of evil for so long that she was beginning to believe it. It was becoming her *own* evil.

The way you see your evil can shift like images in a fog. Your mind changes its perception of evil according to the rational or intuitive channels you select and according to your stage of development at any given time. If you choose to observe evil rationally, for example, you quantify and classify evil. You look upon it "objectively," from a detached perspective, a left-brain perspective. On the other hand, if you choose to look at evil *intuitively* from, say, a Stage Five perspective, you *own* your "evil" processes. You experience the power and life challenge of evil. You allow yourself to be alone in the darkness of your own evil.

Your intuitive mind knows the true nature and power of evil. It knows that to change evil it is necessary to change the holodynes which cause it and to create *new* resonating frequencies. It knows that evil cannot be changed by more evil, nor by isolating yourself among those who are good. Evil finds expression through holodynes which are primitive, immature, undeveloped. To overcome evil, you must enter the heart of the evil, enter the holodynes which give it power, and once there, mature the evil to its fullest life potential. Only in this way can you harvest the power, the life force, of evil transformed.

Is evil an abstraction? To your rational mind, yes, but *not* to your intuitive mind. Your rational mind is the source of most of your conscious fear, avoidance, repulsion, and resistance of evil. Your rational mind cannot understand, much less deal with, evil so it reduces evil to an abstraction and, more often than not, attempts to dismiss it. Now, evil, as the projection of our illusions, is itself an illusion but its *power* is very real. Your *intuitive* mind understands that: it knows how to *transform* the power of evil and release its abundant life force.

Dealing With Evil

In order to truly overcome evil, you must *experience* evil and harvest its potential for life. You no longer *resist* evil; instead, you switch from your rational to your intuitive mind and you *deal* with evil. Your intuitive mind contains the holodynes which empower evil. Once you become in tune with those thought-forms in you which are *your* evil, which have been there since your first fear, which have programmed all your nightmares, and which now program all your evil thoughts and deeds, you can access and explore these holodynes, one by one. Use your I.S.P. to get an intuitive sense of the complete personality of each specific holodyne. If you are visual, allow your intuitive mind to show you the holodyne's image. If you are auditory, you can talk with it. Or you can simply "sense" its presence. But as you do this, it is essential that you feel unconditional love for each holodyne, that you *not* resist its "evil" nature.

What you resist, persists. So, in your imagination, hug each holodyne, befriend it, and love it. If you do not "come to the peace table" with each primitive holodyne, it will remain forever your enemy and will not give you the very information your mind requires in order to overcome your own evil. Once at the peace table, you must find the holodyne's true, positive intent. You must learn to express genuine gratitude for each and every one of your "evil" holodynes and for all the evil they have created in your life. When you allow your mind to accept *both* its life forces *and* its anti-life forces—*both* good *and* evil, you will find the key to peace—to ending your own war with evil, to finally *overcoming* evil.

Once befriended, your "evil" holodynes will reveal to you their positive intent. They will teach you life's true purpose in every evil manifestation. You will learn that nothing is ever done that does not have a positive intent. You will learn that life and all it contains is orchestrated by universal, unconditional love. You will come to see the power of the life force, the power of the love that is expressed in all of life and in every part of your mind, and that springs unconditionally from every evil illusion.

Once you have found the holodyne's positive intent, you can then complete your tracking simply and easily by envisioning the mature image of that positive intent, absorbing the primitive, "evil" holodyne into its mature image, and committing the mature holodyne to accomplish its positive intent in a mature way, and to "be there" for you as a positive life force. You can then invite it and every other mature holodyne to sit at your round table in your mind's place of peace.

In this way, you have conscious access to *all* your mature holodynes, and your mind can call upon their mature processes at any time. In this way, you bring order to your inner world. Your mind's "house" becomes a house of order. And you *potentialize*. You express your new life energy in your actions, your relationships, and in the way you connect with society. You "become" the life force. Your words, your feelings, and your very nature are witness to the fullness of life. In this way—through tracking and potentializing—you *deal* with evil, you *overcome* evil, and you *harvest* its full life potential.

The Lesson of Fear

The particular lessons your mind would learn from evil depend, in large part, upon the stage of development your mind chooses in dealing with evil. Your Stage One mind, for example, may choose mindless nightmares, unnamed terror, fear of dying, or abuse as its challenge. Until you mature their holodynes, you will have no way to handle blind fear, terror, or abuse. Your Stage One primitive holodynes will create an endless, uncontrolled, and often subconscious dance with the forces of self-limitation, the forces of your evil. Such is the case with devil worshipers.

Devil worshipers often "face" their immature holodynes through sacrificial rituals preceded by long and mindless chants. Such rituals allow them to personify the greatest fears of their primitive holodynes, to act out, in ritualistic drama, the terror, fear, and abuse their childhood minds have stored in their subconscious as "evil" images—to see themselves and their fears "on stage." But until they look with *mature* minds upon their primitive Stage One holodynes, which formed in response to their fear, terror, or abuse, they can never truly understand the unconditional love which created their evil in the first place.

When you phase-space evil, you discover only an unconditionally loving "I" would create terror so it could learn fearlessness, or create abuse so that it could learn sensitivity. Only a mind with boundless love for life would make itself dance endlessly with death. The processes of tracking and potentializing allow your mind to release itself from all its fearful dances, to mature up its primitive holodynes, and to fully empower itself for life.

The Lesson of Guilt

Along your path are shadows which challenge you at every stage of development. When you reach your second stage, you become conscious of evil as self-limiting and self-destructive. Your mind becomes an arena for "good and evil" games: your primitive holodynes decide you should be "good" but, instead, are "bad," or have done "bad things," or are "possessed of the Devil." Such conclusions are normal to your immature child's mind, shaped by family and cultural belief systems that resist evil so universally and that condemn childish behavior so righteously. Here your "judger" is given power. Here emotional suffering, temptation, and sin are created. And here guilt is born.

Guilt sets up an endless cycle in which you are "good" until you commit certain acts which are "bad." Then you suffer from "guilt" until you have "paid for your sins" and are free to be "bad" once more. But if you continue to judge evil as "bad," you actually *perpetuate* evil, by *resisting* it, and you will *always* feel guilty. Only by *dealing* with evil, by befriending it, finding its positive intent, and maturing it up can you transform the evil in you into empowered goodness. Your mature mind can then envision the evil within, see the great, unconditional love which motivates *all* evil, and release its abundant life force.

The Lesson of Judgment

When you judge, when you view *any* person, relationship, or system as evil, you encourage evil in *all* persons, relationships, and systems.

Reflect for a moment on the mighty river of judgment and

self-righteous condemnation that flows from so many pulpits. What has all this condemnation, all this self-righteousness ever accomplished, except further divide us, and, in the name of sacred myths about being right, bear constant witness to our collective evil?

But what if you discover some morbid collective evil in an organization to which you belong? Shouldn't you rise up and judge those who are perpetuating such evil? Shouldn't you avoid them or detach yourself from them to find peace from their evil? By your very acts of judging and avoiding, you become *part* of the evil you would judge or avoid.

When you can see beyond your *own* childish evil acts and embrace your *own* true nature once again, you can then see beyond those same childish acts in *others* and reach out to *their* true natures. You can see the futility of judging, you can *stop* perpetuating evil in the name of righteousness, and you can find the perfection in everyone, everywhere. You learn to love completely and, in that complete love, your mind moves *beyond* good and evil, into harmony within itself and with all of life. In such a way can evil be overcome and peace brought to earth.

Good and Evil: Opposite Sides of the Same Coin

In one of my advanced seminars, after reviewing family holo-dynes and exploring family patterns with each participant, I had just enough time to introduce the issues of good and evil. As I entered the conference room the next morning, I was greeted by a close friend of mine, whom I shall call Heather. I was taken aback by the color of Heather's eyes: normally a light brown, they had turned solid *gray*! As I looked again, she said, "I am Lucifer. I am unconditional love." The power of her words and the energy which seemed to emanate from her made the hair stand up on the back of my neck.

Seeing the changed color of Heather's eyes was a surprising new dimension for me, but her changed voice and stance brought back my previous experiences with "evil" holodynes. "Welcome to the party!" I said with a smile. Then, almost as an afterthought, I added, "I will call on you when the time is right."

The group settled down, and another woman, whom I shall call Susan, began to share her experiences of early childhood. Susan told us how she witnessed her mother murder her sister and her father murder her brother. We started to explore the holodynes involved. Susan struggled to gain access. She was so young when the trauma occurred and she had resisted so long, she could not get a clear picture. Suddenly she switched into an "evil" holodyne, which shouted, "I am Lucifer!" Then she became confused as her rational mind refused to own her own evil.

At that point, I looked over at Heather and said, "Now is the time." Heather stood and faced Susan. "I am Lucifer," said Heather. "No. *I* am Lucifer," said Susan. The group laughed. I turned to Susan and said, "Can you allow Heather to act as an alter ego? Can you allow Heather to represent your holodyne of evil? Perhaps that way you can handle it more clearly." Susan agreed.

Susan was able to confront her holodyne of Lucifer with the help of Heather. The image appeared to Susan as a great, red beast with cloven hooves and horns. We then were able to mature the anger, hate, and anguish which she felt from this beast. We got to her fear of death and her sense of loss, and she came up with a divine being of light, whom she called "Jesus." He helped her fulfill the desires she had been trying to express since she was two years old, and the red beast matured into "Jesus."

The process was completed and the group was about to go on to something else when Heather, standing alone in the middle of the room, suddenly said, "I am Lucifer."

The way she said it stopped every sound in the room. "What is it you want?" I asked gently.

"I want you to understand *me!* I want you to appreciate *me!*"

"We want to understand," I suggested. "We love you." That stopped her. She hung her head.

"But you don't *understand!*" she exclaimed.

"What is it we don't understand?"

"You don't understand my sacrifice." She stood with her hands on her hips, her manner defiant.

"Sacrifice is an illusion," I suggested.

"But my sacrifice was as great as Jesus'. I—"

"That's an illusion. Jesus never considered what he did a sacrifice. It was an expression of love." I was phase-spacing, using a Stage Six perspective on the Mind Model.

"What is it you really want?" I asked with real intent.

"I want people to understand my unconditional love. I want them to give me the glory I deserve!" she answered.

"So you want others to see your glory. Can you see it yourself?" She answered she could, so I pursued it a little further. "What does this glory look like?" I asked.

"It looks like the glory of God. I am the Sun of the Morning. I am unconditional love."

"Can you receive the glory of God within you? Can you allow this glory to absorb the very nature of your evil? Can you allow your unconditional love to become the love of God?" I probed, using the tracking process.

"Yes! I can!" she exclaimed. "I see a wonderful light around me. I see two beings standing in the light. They are hand in hand. One is Lucifer and the other is Jesus. I can tell they are in perfect harmony with each other. They are *both* unconditional love! I see a portal opening. They are standing in a stream of pure energy which is flowing into the room. Harmony is pouring forth beyond anything I've ever experienced in my life!" The room was filled with a sense of love that was overpowering: everyone was totally aware, totally in tune.

A woman's voice said, "Is Mary here?" Another woman said,

"I am Mary." The woman representing Mary arose and came to the center of the room. She sat in a chair. Her countenance was indescribable. I asked if she had a message for us. I never have seen such perfect love in another human being. The woman said, "Love my Son." Every person in the room resonated with the love which flowed through her and from her. "Love my Son and take His image into your minds."

Then another woman said, "I have been bad. I want to talk to Mary." "Would that be all right?" I asked. "Mary" nodded and the other came immediately to her and knelt down. "I have been bad! Oh, but I have been bad!" the woman exclaimed. She put her head in "Mary's" lap and "Mary" began to stroke her hair. The love that "Mary" expressed was so calm, so peaceful, so total, that every sin in the room seemed to melt before its acceptance.

"Accept my Son into your heart and love as He loves," said "Mary." And a calm settled upon the woman who had been "bad": she let go of her "Lucifer" holodyne and transformed it into "Jesus" and then became one with the group. We talked with "Mary" for some time. People asked her all kinds of questions, some about what it was like to have been the mother of Jesus, some about how love and acceptance worked. It was interesting to see that "Mary" responded to intuitive, searching questions, but was not interested in supplying rational information. It was as though her goodness were governed by principles which allowed her to speak only to our hearts.

The experience was, for me, the most profound demonstration of the synthesis between good and evil I could imagine. Intuitively, I was phase-spacing the whole thing. I saw how, in the mature love of God, the devil becomes one with the Christ, and the power of evil becomes the power of Christ in every mind. I came to realize that the mature mind sees the devil and the Christ (or God) within itself. The mature mind stops projecting evil onto other people, stops expecting others to serve its primitive images and expectations about goodness, and looks inward for the source of evil and the way to move beyond it, to tap into its life force.

When you phase-space good and evil, you can see how they are opposite sides of the same coin. You can see evil in a new light. Evil is not only the antithesis of life, it is also the opposition required by your mind to orchestrate your fully empowered self. In the Implicate Order, evil has its own duality. There is good in every evil.

Unconditional Love

Once the foundation of unconditional love has been laid, your mind can sense and understand the deep order within every mind and can move *beyond* good and evil. Here you unfold your "I" in all its power. Here you behold the glory you were before you began this life. Here you feel the love we really have for one another. In this new dawn of eternal love, your mature mind responds creatively, constructively to every situation—with harmony and clarity.

When you acknowledge the love you are, you open your heart to the real intent of every evil act, the love which hides in every "evil" heart. When you open your heart to unconditional love, you find the power to mature beyond your own evil.

Your movement into unconditional love creates a mental void for your rational mind. How can you explain to your rational mind the love you are and the love of those who would do evil unto you? You focus on your "I." You use the Mind Model. You track. You potentialize. You phase-space the whole process and your rational mind responds fully. When you "become" the love you are—both intuitively *and* rationally—you witness love in all things.

You can now see evil as it really is, an illusionary shadow cast by your childhood mind upon your life so that you can learn to be in harmony, realize the life power you have always been, and come again into the perfect union of love for every human being.

7

Interpersonal Well-being: On Masculine-Feminine Balance

Because your mind works as a holodynamic unit and your memory contains all the masculine and feminine images you have ever encountered, you have a complete mental map of all the games that males and females play. Within the deeper orders of your mind, your masculine and feminine holodynes are continually at work following your focus, unfolding your potential to love, and creating the patterns by which you relate to the opposite sex. And *you* get to choose at what stage of development you will play. There are always people around who will want to play with you at each stage and they get the signal as soon as you make your choice.

If you are a male, your mind will consciously identify with male characteristics and will tend to portray these characteristics in whatever you do. Subconsciously, however, your mind presents a panorama of paradoxes: your feminine holodynes act as intuitive counterparts to your masculine holodynes, continually promoting their feminine influence. If you, as a male, want to "play hardball," the feminine side of your mind will want to play "softball." Your femininity is continually counterbalancing your masculinity.

If you are a female, your feminine holodynes will usually dominate and make you appear to be what females are to men. You will tend to portray feminine characteristics in whatever you do. And your masculine holodynes will continually counterbalance your feminine holodynes.

This masculine-feminine counterbalancing goes on at whatever stage of development is programmed into your holodynes. Most of this programming has been modeled for you through your family or social bonding experiences. Some programming has been passed on to you genetically. And some has been received by you from parallel dimensions or created especially by you for your own needs. Whatever the source of the programming, your mind has *both* masculine *and* feminine holodynes. Both have power. Both are active in your life. No male can long deny or repress his feminine holodynes; nor can any female, her masculine holodynes.

Masculine-Feminine Games

When the female first learns to relate to others, and particularly to males, she may play at being feminine simply because the male is playing at being masculine. She will play the kind of woman she thinks "he" will admire. If the male responds according to the "rules of the game," he will play the kind of man he thinks "she" will admire. And as he plays harder, she plays softer.

At this stage, the male is usually afraid of his feminine side, so he overcompensates and projects himself as a "macho" man. His feminine holodynes get him to seek softness so that his "macho" male can be controlled: he looks outward, seeking a woman who is soft. His masculine holodynes insist that he remain "macho," so that they will never lose themselves in the softness of their feminine counterparts.

In this game, the female typically admires in the male the masculinity which is *hers,* but which she cannot lay claim to. She cannot claim it because she does not recognize it as her own.

The male typically admires in the female the femininity which is *his,* but which he cannot accept. He does not want his inadequate masculinity absorbed into the softness of his powerful femininity.

Since the male may only love his own femininity in the female, he envies her for her femininity. Since the female may only love her own masculinity in the male, she envies him for his masculinity. Such envy poisons the love dynamic and slowly kills the relationship.

He covets her unattainable femininity and decides to punish her. She covets his unattainable masculinity and decides to punish him. He defends himself by *denying* her femininity, by treating it exactly the way he treats his own. He suppresses it, ignores it, corrects it, and he becomes more aggressively masculine. She treats his masculinity exactly the way she treats her own. She *denies* it, represses it, is frightened and disgusted by it, and she becomes more submissively feminine. Their masculine-feminine balance grows more and more lopsided. Both become less and less of what they want to be and what they really are.

The game is a total shutdown. It is the death of love. If he were to yield his power to her, he would suffer a loss of integrity, a loss of his masculinity. If she were to accept his power, acquire his masculinity, her femininity would be devoured in the process.

Changing the Game

The only way to win at this game is to *change the game.* And to do this, you must first phase-space your relationship. You must look at what is really going on and decide to invest in the game of life. It only takes one of you to find the pathway back to life and to mature love. It works better, faster if both of you decide, both look and both travel the journey together. But one of you can do it—if that one *decides* to do it.

Once you have decided to go for *life*, once you have *chosen* to love with a fully potentialized masculine-feminine balance, your next step is to access the hidden holodynes which empower the game you are playing. If you are a female, these would be your masculine holodynes; if you are a male, your feminine holodynes.

If it has been a long war, the female has buried her masculinity beneath the dead and dying "bodies" (past memories) of her soft, sensitive, feminine holodynes. To give up the game, she must see that all her battles were "perfect," that, with them, her mind was teaching her the value of her softness, the value of her intuitive vitality, of her love of life. She must deal with all her psychic monuments to death, and mature them up before she can find her strength, her assertive masculine self.

If the male is to change the game, he must allow his hard masculine self to *accept* his soft feminine self. He must see that all his male-female battles were, in fact, schools of learning, and come to value the lessons he learned detaching from his own sensitivity and denying the tender and receptive parts of his own, true nature. He must face his fear of vulnerability, and step *beyond* it, before he can find the hidden strength of his sensitive, open feminine self.

This is why so many of us fail to free ourselves. We cannot let go of our attachment to our miserable memories—our "war memorials" to our failed masculine-feminine potentiality.

Among the female's "war memorials" are memories of unfulfilled relationships with males throughout her lifetime: with her father, her brothers, her male friends, her husband; and among the male's are memories of incomplete relationships with the females of his life: with his mother, his sisters, his female friends, his wife.

You both cling to the love-hate game you have been playing in your minds, among your holodynes, and therefore in your life together.

Only when you discover the game you are playing, only when you see the dance of death going on in your relationship, and *own* your part in the dance, can you begin your passage to maturity. You start, at Stage One, by *recognizing* your *other* side: as a male, your feminine side; as a female, your masculine side.

At Stage Two, you learn to *access* those holodynes which have been orchestrating the game, to talk to them, to *befriend* them, and to appreciate them for the positive intent of what they have done. It is during this stage of development that the female comes to *know* her masculine self, to understand that, by seeking the masculine in the male, she was really seeking the male in herself. She may reflect upon all she has wanted in a male and see it in herself. Or she may intuitively envision the male within her, and recognize in his full potential her own. She learns to befriend, love, and accept her masculine self—to bond freely, intimately, and maturely within herself. And the opposite, of course, applies to the male. If he is to change the game, he must *own* the female within himself.

At Stage Three, your relationship takes on causal potency in the inner world of your holodynamic plane. You choose to care maturely, you *commit* to live and love in a new, *balanced*, masculine-feminine way, and a new set of holodynes is born. You emerge whole. You no longer "need" another person to feel complete. You *are* complete within yourself and, in that completeness, you are ready to bond maturely.

Only when you are *free* to love, when you don't *have* to love, can you get in touch with your *innate* ability to love and learn to love maturely. To release your love potential, you must first go to your place of peace, access all the immature holodynes—both masculine and feminine—that are holding you back, keeping you in your old love-hate patterns, and mature them to their fullest potential. You must *track* and *potentialize*.

A word of caution. How you communicate with your holodynes at your place of peace will most likely be affected by how you have experienced intimacy with your family and friends. At Stage Three, your mind will often use such models of intimacy to govern how your masculine and feminine holodynes communicate with one another. If males and females in your family and among your friends communicate in ways that are open, deep, mature, and loving, your round table will become an open, effective meeting place for your mature masculine and feminine holodynes, an accessible, deeply satisfying resource for daily guidance as you interact with the opposite sex. You will be able to experience love more profoundly and to share a degree of intimacy not often found in our modern society.

But if, on the other hand, your models of intimacy reflect distrust, animosity, and war, you can get lost just setting up your round table. Sometimes your intuitive guides will come, but once there, they won't talk. Sometimes they will disappear just when you need them the most. Under these circumstances, you may find yourself blaming your family, even hating them. You may want to disown them, withdraw, and seek other models. It's easy to blame your family. It's also a distraction. Your family is *only the way you see them.* They have "been there" for you: they have acted out, modeled for you, all the issues your mind wanted to learn. Therefore, don't "disconnect" from them. Instead, *own* the reality of "I *am* my family."

So, acknowledge that the issues you have with your family (or with any of your intimate friends) are really *your* issues. Then and only then are you ready for the "real work" of Stage Three. Then and only then can you access the holodynes which have created the issues, listen to them, and mature them. The secret here is: align your masculine and feminine holodynes with your Full

Potential Self. Realize that only when you are truly connected to your *own* core are you capable of truly connecting with another person, core to core, soul to soul.

Create an internal environment of peace, of communication, cooperation, correlation, and dedication. Invite your mature holodynes to sit at your round table, to have an equal voice, to be heard, to act maturely, responsibly, and lovingly toward your immature holodynes so that *all* your holodynes can work together, love one another, and help you unconditionally love others. This open, accepting, loving approach prepares you for Stage Four.

At Stage Four, you must engage your cultural beliefs and experiences. How you organize your mind at this stage will be how you organize your external world, and will reflect itself in how you relate to others. If, for example, you believe that your culture creates the differences between men and women, you may begin to focus on specific cultural issues in an attempt to deal *externally* with your *internal* dynamics. At this point, you may find a certain "comfort" in blaming society for all its—and your—problems. You may sift through the endless variety of "specialized" roles within society: "mother-nurturer," "father-discipliner," "male-provider," "female-homemaker," and the like. "*Society* keeps men and women different," your Stage Four downdraft holodynes may say.

By focusing upon society's part in keeping masculinity and femininity separate, your mind is beginning, in an indirect way, to look at the issues which it faces in its own internal community. If you do not *own* these issues as *your* issues, and find a way to create unity and harmony within the society of your holodynes, you may find yourself endlessly engaged with social issues—trapped once more in the "duty-obligation" games which limit your love and which never change society. To change the society outside you, you must first change the society *within* you. How you participate there, in your internal world, among your holodynes, will be directly reflected in how you participate in the world around you.

By blaming society for sexual differences, you *project* onto it *your own* masculine-feminine games. Your mind sees *outside* itself what it hasn't overcome *inside* itself, and thus does not have to face its own dysfunction. As long as you are "working on the problem" by "seeing it in others," you don't have to "work on it" from within. Thus your mind reinforces all your primitive illusions about males and females. You re-create the polarizations, over and over again. And you pass them on from one generation to the next.

Now, in such a frame of mind, you can see how often and how powerfully society reinforces sexual polarizations. When you find yourself locked into this kind of thinking, phase-space it. Look at it through the Mind Model. Stop expending your energy in games and start using some of it to identify each element within the culture which "feeds" into the games. Be careful not to get caught up in the futile game of classifying and quantifying all the social roles, scripts, expectations, teachings, rituals, and beliefs which separate males and females. You can endlessly divide and sift through the data for "empirical" evidence that men and women are, in fact, "justifiably" separate. Instead, *face* the responsibility for intuitive and internal change—*own* the "researcher" within you who helps you *avoid* your real issues, and you will see that the external issues you concern yourself about are really your internal issues. And you can then move on to the next stage.

Your unresolved masculine-feminine issues can be very deep and intricate. In this regard, the traces of all your ancestors are still contained within the holodynes of your mind. Every unresolved male-female issue, every male-female conflict from past, present, future, and parallel worlds, resides within your memory and is available to you as part of your holodynamic mind. From your ancestors, from the precise modeling of your family and culture, your mind has drawn a map of *every* male-female game, no matter how subtle or complex.

Take, for example, the case of a mother of several children who was sexually shut down. As she began to explore her holodynes, she discovered deep, unresolved issues between her father and herself. He had had incestuous relations with her for about fourteen years of her life and, even though this was long ago, she still held deep feelings of rage, guilt, and inadequacy about her own sexuality. By learning how to track and potentialize, she was able to mature her immature masculine and feminine holodynes and her sexuality became empowered once again. She and her husband were delighted.

As you emerge to your fifth stage of development, you are ready to face all your immature masculine and feminine holodynes, whether from family, group, culture, ancestors, or parallel lives. You can rejoice in the perfection of these contrasting, conflicting wonders of life. You take the old battles between your masculine and feminine selves and see them for their real intent. You openly face your immature holodynes, accept them for their positive intent, love them, mature them up, and invite them to join with your mature self to find their—and your—fulfillment. You *"become"* your masculine-feminine balance. You recognize the value of church, school, and all cultural systems, but you own your primary responsibility to make your own house a house of order.

With your new, internal integrity, you can view anyone who wishes to continue the old masculine-feminine games from an entirely different perspective. You can reach deep into the holodynamic plane of any mind. You can sense the Full Potential Self of each person and reflect the potential which awaits unfolding within him or her. Using I.S.P., you can invite each person's Full Potential Self to meet regularly with yours in your place of peace. You can sensitize yourself to the real potential between you and begin to act in accordance with this potential. In this way, person to person, soul to soul, you become a witness to the unfolding of our fullest potential as men and women within a larger society. You "become" that potential.

In the sixth and final stage of your development, you come to know that your masculine-feminine balance is a wonderful reflection of the underlying harmony and symmetry of the universe. You *extend* the power and love of that balance to all people, all life, to the universe itself. Not only are you attuned to the potential within others, but as you extend yourself fully to them with your love, you become empowered as a man or woman.

A 6,000-year-old Case History

In order to see how transgenerational patterns and parallel worlds can impact your present life, consider the case of Darlene. Her presenting symptom was sexual rage. She felt an uncontrollable anger toward her husband when he wanted to have sexual relations. She was a licensed social worker, a psychotherapist, and a sexual counselor. She could not understand her rage.

I introduced her to the multi-dimensional aspects of her mind, showed her how to use the Mind Model, and she began to access her holodynes. From her place of peace, she was able to access her Full Potential Self and enlist it to be her guide. She then asked her Full Potential Self to show her the holodynes which were causing her sexual rage.

The first thing she saw was herself at four years of age being sexually molested and forced into sodomy by an older man toward whom she felt great affection. As she "became" the four-year-old, she was overcome by confusion: her feelings of affection became mixed with growing rage.

We asked the holodyne of her little four-year-old to talk to us and to let us know what she really wanted. She replied, "Love and peace." She was being abused and she wanted it to stop—

without losing her love for the older man. We called forth a mature holodyne of love and peace to help her four-year-old. After some detailed training on how to love, what sexual abuse meant, what the older man really wanted, and how to achieve mature love in life, her four-year-old was absorbed into a mature holodyne of love and Darlene's rage subsided.

Several months went by. Darlene requested advanced training: some "unresolved issues" were still "haunting" her. She prepared for a "reliving," wherein she would explore the holodynes of her ancestors in parallel worlds. In drawing up her genogram (see Appendix, p. 208), she noticed a relatively large number of divorces in her family, and became aware of a great deal of emotional energy surrounding male-female games.

Immediately into her "reliving," she saw a beautiful temple. It was very ancient and in it a group of young men, all dressed in white, were sitting upon a white marble floor receiving instruction. She was the only girl and had come dressed like a boy. Suddenly she was looking down at a dark red stain. She was startled to discover she was having her first menstruation.

Frantically, she tried to clean up the blood but it wouldn't come off the marble. She tried to flee, but one of the priests saw the blood and held her captive while the high priest of the temple was called to question her. A form of trial took place. They demanded to know what she, a woman, was doing in a temple *forbidden* to women.

She heard herself explain to these ancient priests that she had disguised herself as a boy and had entered the training so she could find out the mystery of how to become one with God. She wanted to share this with the women so *they* could become one with God also.

At this admission, pandemonium broke out in the temple. The men were outraged at such a violation. They seized her, put her on the altar of the temple, and cut out her female organs in a brutal operation. She stayed alive several days in unbearable agony. No one was allowed to help her in any way. She died with terrible torment in her soul. The men of the temple were more determined than ever that women would have no role in the affairs of God. She was able to place the date of the incident at about 4,000 B.C.

Her guide, her Full Potential Self, then took her through a series of experiences which demonstrated how the attitudes of the people involved became ingrained in the male-female relationships of their descendants. How the men always took care of the affairs of God and the Church. How the women were not allowed to intervene in any way. It became the custom of her ancestors.

During the crusades, the male-female scenario took another dramatic turn. She saw herself as a princess being escorted by four of her chosen guards. She was about fifteen and very beautiful. Her guards were to leave for the crusades the next morning for the adventure of their lives. She was riding through a forest, with her guards slightly before and behind her, and feeling proud of them and their coming adventure when, suddenly, they drew up and surrounded her. They dismounted and, holding her horse, teased her to get off.

Still trusting them and partly enjoying their antics, she dismounted only to be forced to the ground and gang raped by them. They abandoned her, to find what shelter she could in the falling night, and she never saw them again.

The terror and hurt she experienced were never resolved. The generations which followed carried the scars of the broken trust and the "beastly passion" of the men. The women built up their defenses against their vulnerability to men and learned to play a series of games with them. These included flirtation and coquetry— "I will promise you everything but will give you nothing"—and were fortified by other brutal episodes over the generations, as the men retaliated with games of their own.

Darlene's Full Potential Self explained that the sexual abuse she had received from the older, trusted male in her family was empowered by these family traditions. Men had learned to use and abuse women sexually, to perpetuate their primitive beliefs through incest, as well as through other hurtful games played with women. She was born into such a system, her Full Potential Self informed her, so that she could help purge her entire family of the poison within its roots.

In order to release her family from these traditions, she had to "relive" each experience, talk to all the holodynes involved, and mature them to their fullest potential. We took the young girl dressed as a novice priest and introduced her to a new guide, a mature woman who was "one with God." Darlene's potential for "oneness with God" unfolded. A whole series of religious issues came to the forefront, and her new guide explained each one to her. The guide showed Darlene where and why each one originated, and how men's insecurities had demanded a "specialness" from the women, which was part of the whole illusion. She unveiled Darlene's own "specialness" as a woman and her part in the illusionary game with men.

The guide then unfolded the nature of Darlene's own womanhood, the various masculine and feminine holodynes involved, and how the balance between the two was part of the Implicate Order of nature. No "specialness" was required since the combination was innate within her and provided the fountain of life. In the process, all the priests and victimized women were absorbed by her new holodynes of mature masculine-feminine balance. Darlene's issues with authority figures, especially males, disappeared.

As we applied this same process to Darlene's experience with the young soldiers about to embark on their crusade, she saw how *their* traditions had spurred them on to "one last adventure" with her. She felt their love, their shame, and their resolute determination to "become men." Her guide then took her to a dimension before time, to her "place of planning," wherein she saw each of these young men at their full potential. They were radiant beings and had a total, unconditional love for her. She observed how she and they planned the "test of love" for her, her rape, her abandonment, her anguish.

She wept openly as she "experienced" the lessons her mind would learn from these encounters, and how, down through the generations, a resonating energy field would carry these same lessons into the lives of hundreds of her family. She saw how all had agreed before time to these lessons. She understood that there was nothing to forgive. The unfolding of thousands of years of traditions was agreed upon so that the lessons could be learned. There were no victims, no perpetrators of diabolical deeds. Only totally loving people acting out circumstances to test the un-conditional love between men and women.

Her guide then showed her the man who molested and abused her when she was four years old. He was her grandfather, whom she loved dearly, more than any other man. She saw him at his full potential and saw the agreement made between him and her in the unmanifest dimension of her place of planning. The lessons she would learn, of pain, confusion, anguish, rage, sexual inhibitions, and distrust of men were all made possible by their agreement. The agreement, she reported, was one of total love. "Only someone who totally loved me could give up so much of himself to teach me such a devastating and all-encompassing lesson." She wept.

Darlene had completely transformed her rage. In doing so, she felt she had also released her family from the love-hate games its men and women had played with each other down through the generations. Her sexuality blossomed. She became a warm and loving woman, and a wonderful wife to her husband.

She reports she can now help her clients in ways she never

dreamed of before. And since her "reliving," she is still learning about herself, getting more and deeper insights into her own masculine-feminine balance, which she can share with others.

Her husband, Dave, a prominent businessman, phoned me two days after she came back from her training. "What happened to her?" he asked. "She's a completely new person! It's wonderful to see such a transformation. How in the world did it happen?" I told him. He took the training and had a similar experience.

Dave's family traditions complemented Darlene's almost perfectly, although they came down through a different family line. When he experienced his "reliving," Dave found himself in a series of wars and was a witness to terrible devastation: the burning of cities, the rape of women, the slaughter or imprisonment of entire populations. He had laid a defensive-aggressive blanket on his own masculine-feminine balance. We matured these violent episodes into the lessons of their positive intent and Dave's mind began to release itself. He began to move *beyond* the wars. He came to be at peace within himself.

Darlene was delighted. Dave was able to come into a new dimension of harmony and fulfillment with her that has allowed them both to become more active in their community. Once balance is established within one mind, its resonating energy field influences everyone and everything around it. Thus Darlene's balance inspired her husband's. And his inspired her. A new dimension of intimacy and love, of understanding and cooperation unfolded within their relationship. The balance in their relationship inspired others around them to experience their own personal "balancing."

Masculine-Feminine Projections

You can see the "other" side of your masculine-feminine self in your *projections*: if you are a male, you tend to project your feminine self onto females; if you are a female, you tend to project your masculine self onto males. Projection works like a movie. The actual pictures you see on the picture screen are not on the screen but in the projector. In this same manner, you project onto *others* those things you see in *yourself* but do not wish to face directly.

Take a few minutes and make a list of all the characteristics you can't stand in other people, especially in the other sex. Then, when your list is complete, read it back with this thought in mind: *"All these characteristics are holodynes within me that I project onto others."* It's a great reality check and it will tell you very clearly what your subconscious mind is working on, especially what issues you may have with your "other" side.

Your images of the other sex, whether negative or ideal, are usually projections of your immature feminine holodynes, if you are a male, or of your immature masculine holodynes, if you are a female. For how you see the *other* sex is literally the consequence of how you perceive the "other" holodynes within *yourself*.

When you love maturely, you accept the other person for who she or he really is, as a human being, as a part of your life. You freely choose to love the other just as she or he is. You do not need to overcompensate with ideal images because you do not suffer from inadequacy. You have matured your old inadequacy holodynes. You have passed beyond images and projections into the real world of living relationships.

As a mature lover, you know that projections and images are a perfectly natural part of every relationship. You understand the process of projection and can move beyond it into mature commitment, disciplined love, trust, and balance. You see in each projection an opportunity for self-discovery, a new view of your own masculine-feminine holodynes, a way of knowing your heart more deeply. You treat each opportunity as an invitation for bonding, and balancing, a further adventure on the path of intimacy.

Toward Mature Bonding

You have, within your intuitive mind, the innate ability to love and to express your love in a mature relationship. But to love maturely, you must *un*learn most, if not all, of what you've learned about love from your family, your friends, your peers, and society as a whole. You must *let go* of your love-hate games and *potentialize* your immature masculine-feminine holodyne—at *every* stage of development.

To most people, "mature bonding" means "marriage." Let's look, then, at how your attitude toward marriage and your capacity for mature bonding *evolve* from stage to stage.

BONDING AT STAGE ONE

At Stage One downdraft, you see marriage as a "dead" institution, responsible for passing on the "sins" of generations gone by. You gather endless data on marital failures, you focus upon rising divorce statistics, you notice how many couples married 20, 25, 30 years are going their separate ways. You believe there are no happy marriages. Armies of psychologists and marriage counselors, you point out, have not been able to stem the rising tide of divorce. "Marriage is archaic," you say.

And why do you focus upon the *death* of marriage? What would your mind have you learn? Rationally, you criticize marriage, you bemoan its death because *intuitively* you seek the keys to life, to a happy relationship, to mature bonding. You pick apart dead and dying relationships in order to discover the nature of their dis-ease. Why? So you can cure the disease within your *own* relationship—within *yourself*.

By realizing that the "disease" is *yours*, you place your mind in the *updraft* of Stage One: you move toward a *"life"* perspective about marriage and *away* from your rigid, immobilizing "death" perspective.

At Stage One, change occurs when you decide to look *within* and accept responsibility for *your* part in the game you are playing—when you decide to shed the protection of your comfort zone beliefs and stand naked before your truth. You phase-space your bonding games and discover you *are* your problems. You *created* your problems as a challenge. You *created* your marriage as an opportunity for intimacy. And because you are in the flow of life, you can grow and you can *change*. You can choose to put life into your marriage, to unfold within it your hidden potential for mature bonding.

BONDING AT STAGE TWO

At Stage Two downdraft, you see marriage as a battleground, a constant struggle. "I am married to a psychopath!" you say. *Other* people have happy marriages: if only your *spouse* would do this or that, then you would be happy. Such thoughts can soon carry you back to your old Stage One shut-down if you are not prepared to see marriage as an opportunity to rub up against your most intimate self and to realize that your partner simply cannot "make" you happy.

Once you experience abuse and lovelessness you come to appreciate the meaning of love and to understand that abuse and lovelessness are *your* illusions. They have nothing to do with your partner. Your struggles are opportunities for you to develop faith in your own power of love, to learn about yourself, and to move *beyond* your old myths.

No one can *provide* you with harmony. If your own mind is at war within itself, no "external" peace can touch you. You might marry a saint but your war would continue endlessly and you would come to hate your saint for constantly reminding you that you hadn't "made it" yet. Your lesson, at this stage of development is that *you* must work out your *own* harmony, your *own* salvation: happiness is the natural by-product of a peaceful, harmonious mind. You must decide to care enough to *overcome* your self-defeating behaviors, to confront, befriend, and mature up those holodynes which are orchestrating chaos in your relationship with your spouse.

At Stage Two, change occurs when you intuitively befriend your immature holodynes. You come to realize that your mind created these holodynes when you were young and could not see the larger picture. You come to realize that each immature holodyne has a positive intent—to help you learn.

Your immature holodynes teach by contrast. They orchestrate lovelessness to teach you love, panic and anxiety to teach you patience, deprivation to teach you abundance. Your hard-active-dominating-forceful masculine holodynes teach you sensitivity, openness, and vulnerability. Your soft-passive-avoiding-submissive feminine holodynes teach you assertiveness, clarity, and strength.

Each "primitive" holodyne is part of a wonderful life process of growth: as you intuitively experience your holodynes, you come to appreciate all they have taught you. And once you see what they really have been doing, once you focus on their *real intent*, and stay focused, you can bond with them, balance them, and align them with your Full Potential Self. Only when you learn to bond within yourself can you bond with another person. Only then can you move on to the third stage of development.

BONDING AT STAGE THREE

At this stage, your mind leaves behind its rigid rational expectations and you sense the dynamic potential of bonding in your marriage relationship. Acceptance of your masculine-feminine self at Stage Two prepares the way for unconditional acceptance of your spouse at Stage Three. You commit to your relationship and, as you focus upon it and put your life energy into it, it comes alive. It gains its own causal potency.

You can phase-space your marriage relationship as a living entity, so as to see clearly what gives it life or causes it to die, what increases or diminishes it, limits or potentializes it. A healthy, living, empowered relationship lets you freely unfold your potential for mature bonding. A sick, dying, disempowered relationship, one that requires your spouse to match your images or to constantly please you, draws you and your spouse into endless cycles of frustration and quarreling.

Your ability to bond in a relationship depends upon how well you have bonded the holodynes of your mind and how well you can express this internal bonding in your life. For example, do the holodynes of your mind openly and freely communicate with each other? Does your Full Potential Self align your holodynes so that you can act with real integrity? Only when you have such integrity can you be fully committed to your partner. Only then can you bond maturely.

If your commitment to one another is to learn together, to walk life's path together through all kinds of weather, you naturally bond through friendship, through *shared* responsibility for the relationship. Neither one of you has to take responsibility for the other because each of you is whole, capable, and confident in his or her own abilities. Your relationship is healthy because each of you can respond to the other, make room for pain or immature behavior, and never lose track of the real potential within the other and within the relationship.

If, on the other hand, you downdraft and you join with another because of your "need" to find protection, completion, or fulfillment, you naturally *bind* yourself to your partner and continue to feed your old, primitive childhood dependency holodynes. You avoid personal growth and responsibility; you refuse to explore feelings, anticipate behavior, or look at consequences; you seek stability through compliance, denial of self, conformity, and indifference; and you expend your energy trying to control others through guilt, fear, and judgmentalness.

Bonding is a matter of choice, a natural part of the Implicate Order of growth. Once you choose to align yourself with your potential to bond, and with your Full Potential Self, you can align your actions so that your behavior reflects this hidden potential. You can develop the rapport, mutual respect, and genuine friendship, through shared inner selves, that will lead naturally to real intimacy.

Even if you both begin your relationship by being caught up in blind, euphoric passion, you can bond. Even if you both come from completely different family and cultural backgrounds and find, "after the honeymoon is over," that you can't stand one another, can't eat what the other eats, and can't even sleep on the same schedule, you can still bond. You can learn to harvest the lessons that come from such cross-pollination and use your differences to form even stronger bonds. Unconditional love shines brightest where there is no reason to love. It is the power that pulls us all beyond our limited thinking and that opens our fullest potential into this world. And it can only develop when you align yourself with your fullest potential and with the fullest potential of your relationship.

Only within an intimate, committed relationship can your deeper, hidden holodynes be revealed. It takes years of interaction for your spouse to learn to reflect them clearly. Thus, all the things you resent about your spouse are reflections of your *own* immature holodynes.

All the pain, distancing games, downdraft thinking, fighting, boredom, passivity, loneliness, and lack of intimacy are reflections of your own holodynes. All your victim-pleaser roles, image matching, manipulating, and disconnecting are part of the game you want to play in some part of your mind. Change your mind and you change the game. Commit to bond intimately with another person, learn bonding skills, and genuine intimacy will follow naturally. You will know intuitively how to keep intimacy alive, healthy, and growing. And as your intimacy grows so, too, will your capacity to bond.

Just as union within your mind creates union within your marriage, so union within your marriage creates union with other people as well, for once you truly accept yourself, you can truly accept your spouse and then, surely, you can truly accept others. Mature bonding within a larger social group is possible only when each person adds a strong link to the whole chain. Once you and your spouse have learned to bond maturely, each of you can form alliances that can withstand group pressure, each of you can adhere to group objectives and cooperate with trust, teamwork, and genuine collective conviction.

When you *fail* to mature your holodynes, you secretly subscribe to your insecure, inadequate, and self-defeating definitions of yourself and you will tend to *fuse* with others of like mind—to *melt into* the group identity. Here, you will likely maintain a life of personal invisibility, forever spouting group dogma, conforming to group rules and regulations, without personal insight, personal responsibility, or personal inspiration. From your blind faith, fear, inadequacy, and guilt, you speak for a "cause" or for "God" or for "the company" by giving vent to the immature holodynes that rule your life at every stage of your development. Your life, your marriage, and your group activities remain hollow, empty, lacking the fire of personal conviction, and barren of living spirit.

Only when the fibers of individual integrity are joined together in mature friendship can healthy group bonding result. Such a group handles masculine-feminine issues in mature, effective ways. Sexual discrimination and harassment seldom occur because the group dynamic does not encourage them. By phase-spacing abuse of any kind, each of you can identify immature behavior, find its positive intent, and then mature the holodynes involved. Within your group, you can change even deep-seated prejudices, built into the male-female games of our society. Personal vitality, individual creativity, self-discipline, respect for relationships, genuine cooperation, collective purpose, and a willingness to tackle problems, no matter how complex, are all part of healthy group bonding.

BONDING AT STAGE FIVE

When you bond maturely, you learn that there is no such thing as *partial* love. When you choose love, love chooses *everyone*. Love conquers all, loves all. When you care enough to mature up your immature holodynes and to establish mature love within yourself, you can maturely love your spouse and, by so loving, come to love all humanity. Love is thus unfolded. It becomes a living force, it takes on a potency of its own.

As a therapist, I found that any aspect of marriage, intimacy, or bonding could be explored and handled more effectively by phase-spacing it and viewing it from the many dimensions of the Mind Model. People who were too left-brained, emotionally locked up, sexually shut down, incongruent, unethical, or hypocritical could often be helped once they understood the natural stages of development and applied these to their thinking and to their lives. Whether they were secretly gambling, drinking, or taking drugs, whether they were cheating on the side or leading double lives, they could see more clearly the choices they had: to align themselves with their fullest potential—to develop internal integrity—or not, to access their immature holodynes and mature them up or live in constant conflict within, and without. It boiled down to one choice: to be true to themselves or not. Once you are true to yourself, being true to others follows naturally, and your integrity develops a power of its own.

The way your mind encounters the self-empowering nature of love and integrity is to think about them in terms of guiding principles. To your rational mind, love and integrity, like faith, hope, and charity, are abstract principles, but, to your intuitive mind, they are streams of empowered thought, guiding channels which direct the flow of your life energy and which determine how you think, feel, and act.

Once your mind is free to explore its guiding principles both rationally and intuitively, you reach a new level of awareness. You realize that *you* created all the games you have been playing in your life. You realize that every game, every pretense, every hypocritical, unethical, or unscrupulous act, whether done *by* you or *to* you, had a lesson for you to learn. And you come to see the lessons of love and integrity clearly. The entire "echo-system" unfolds the love and integrity which are deeply embedded in the fabric of your Full Potential Self, and which become manifest when you *own* the love your *are*. You "become" love, and you "become" integrity. No one can take them away from you: you *are* them.

The qualities of bonding become *your* qualities. You bond sensitively, openly, honestly, lovingly, with awareness, deep appreciation, and integrity because you *are* these things. You stop playing your old games because you have better things to do. Your principled mind empowers new dimensions to your life. You are now able to universalize your bonding.

BONDING AT STAGE SIX

Intimacy is a prelude to universal oneness: the fullest unfolding of the Implicate Order which connects all things. At Stage Six, you can see in the masculine-feminine games of your mind—of individuals, relationship, and groups, in every arena of the human interaction throughout history—the unfolding of mature bonding. You can see a new, dynamic system of personal and social transformation begin to take shape.

When you look at humanity *holodynamically*, masculinity and femininity become "subtle attractors" in the quantum field of history, and bonding patterns between these polarized attractors, major determinants of historical trends. As historian Riane Eisler points out, male-dominated hierarchies of power typically seek to rule by force, which they can do only by eliminating female influences from all governing systems—political, economic, religious, and educational. This leaves them free to create endless wars. Time after time, feminine influence in society emerges to challenge male-dominated models of government. And as this influence makes itself felt in Cretan, Roman, European, and Western history, wars grow fewer and cooperative forms of government evolve, where equal consideration and partnership take the place of dominant-submissive rule. Feminine gods ascend, women gain economic and political power and show much more leadership in political affairs. During these periods, education and cross-cultural bonding flourish.

But, according to Eisler, history shows wave after wave of male suppression of such developments—a seesawing, back-and-forth motion between male and female influences down through the ages. Time after time, such polarizations have occurred in human history. Nazi Germany, Stalin's Russia, and even China's latest suppression of democratic reform reflect the rule-by-force, male-dominant mentality. This unbalanced mentality must be rooted out if we are to survive, if life is to continue on this planet.

Rooting out the male-dominant mentality begins by understanding the holodynamic nature of bonding. It begins with you—at the first level of organization of your mind—the level of holodynes. You hold the key to transforming this world from an unbalanced, rule-by-force mentality to a balanced, partnership mentality. Conflict resolution begins among the holodynes of your mind. If you would have peace in your outer world, you cannot repress, dominate, or "rule over" any of your holodynes: they are the "people" of your inner world. How you treat them resonates outward into your personality. If you repress, sublimate, or dismiss your more sensitive feminine holodynes, or allow your masculine holodynes to dominate them in any way, your internal chaos will create chaos in the whole external field. When you rule your holodynes by force, you naturally set the stage for rule by force in society. Each of us helps establish the basic patterns by which our collective mentality functions. How you relate to yourself, among your holodynes, becomes a subtle attractor in the whole field of human interaction.

It is my experience that holodynes hold the key to all human behavior. Bonding patterns among the holodynes of your mind determine the patterns of your personality. The patterns of your personality determine how you will bond intimately with others. Patterns of intimate bonding determine how groups, and even nations, will relate to one another and thus the well-being of the world itself.

You get to create the kind of world you want within your mind. And by doing so, and then *extending* yourself, you help create the kind of world you want *around* you.

8

Social Well-being: On Teaming Up

Success means more than just making money, having equip-ment, making the sale. Success is more than creating a work of art, having the right home, knowing the right people. Success is also having physical energy and health, being mentally alert, having emotional strength and stability, and having friends you can trust and be open with. Success is doing what you were *meant* to do: acquiring real skill, having confidence enough to *know* something, belonging to a team, and being able to perform well. When you are truly successful, you have a sense of integrity, a feeling about you that says you are *real*, you don't have to put on a show or impress anyone. Real success means being so attuned, so capable of con-tributing, so empowered and knowing, that you are at one with your environment and in harmony with your team.

In other words, when you are successful, you unfold your enfolded potential to work as part of a *team*. You have invested wisely and consistently in your "updraft" bank account and are living off the interest. You have trained your mind to potentialize every negative holodyne from your past: you have tracked and matured all those potential saboteurs that would downdraft the dynamics of your life. You are fully in tune with your Full Potential Self and thus are free to work with others in such a way that you can impact the Quantum Wave and *get what you want*.

When you are successful, you choose to put your life energy into what you do, to unfold your "I" in your professional and social life. You develop deep and honest friendships and associate with groups of people who are like-minded. You create a positive resonating energy field, a reservoir of life energy which supports your success.

Six Steps Toward Building a Team

There are six stages of development. You can make deposits toward team building at each stage or you can make withdrawals. It's entirely up to you. At each stage, once you choose, you activate certain holodynes, which create resonating frequencies which impact the Quantum Wave. Your Full Potential Self takes your choice, checks out the boundary conditions, and you get the consequences. What this means is, "Whatsoever a man thinketh in his heart, so is he." If you want to build a team, you must make deposits into your team-building bank account. You must make the choices which promote team success for you and your teammates. Here are six steps you can take to build up your team power.

1. FOCUS CLEARLY

Focus means more than having a general picture in mind. Focus means clearing away *all* distractions, both external and internal, and using *all* your senses, both rational and intuitive, to zero in on what you and your team want. You must *choose clearly* what you want and then stand in the vortex of its resonating energy field and *keep* your focus clear.

Here are some ways to get your focus clear and keep it clear. First, think things out rationally with your teammates. Take a good look at the "real" world and do a "check and balance." Make sure what you want is reasonable and not likely to be cancelled by the boundary conditions of this plane. Then, once you and your team have agreed upon a reasonable objective, go to your place of peace, sit at your round table, and using *all* your senses, imagine what it would be like if you had *already* reached your objective. What would you be doing? How would you be feeling? What changes would have taken place within your life? Explore your circumstances in detail. Get a real feel for what reaching "point B" would be like.

Once your objective is clearly in focus, you must invest your life energy into *becoming* the success you have imagined yourself to be. Begin to *act* upon that success in every detail. Develop team power among the holodynes within your mind. Choose special intuitive guides to help you achieve your objective. Do everything you can to strengthen its resonating energy field and thus naturally move you and your teammates toward its fulfillment.

Don't let your old, comfortable family and cultural beliefs thwart your investment. Be aware that, as soon as you stand in the vortex of your new resonating energy field, your family, your friends, and your teammates will sense it and, often without knowing it, will do everything they can to *keep* you where you *were*—in *their* old comfort zones. They love you. They want to keep you close to them. They want you to stay the same. They want you to be "comfortable." They will act out for you, mirror back to you, and project onto you, any immature holodynes which you have not yet tracked regarding your success.

So thank them. Love them. Accept the things they do and say that block your success as gifts, as signals that you have work to do on your own holodynamic plane. If they blame you for their troubles, accept the blame and see what you can do to improve yourself. Thank them for their insight. Then reflect back to them their real intent. Get them to imagine what it would be like if everything was just the way they wanted it. Negotiate to see if they can join in your effort. Work together on getting what everyone wants. Enjoy one another. Combine efforts. Cooperate. Be friends.

Then, in the privacy of your own mind, access those holodynes on your holodynamic plane which triggered such criticism. Go deeper. Relive the feelings you had when you were being blamed. Use your I.S.P. and access each feeling. Find its intent. Track it. Potentialize it. Clear your own mind. Keep your focus. Keep the resonating frequencies clear. This will maximize your contribution to the team effort and your possibilities for success.

2. UNFOLD YOUR PERSONAL POTENTIAL

Unfolding your personal potential means aligning your conscious self with your Full Potential Self, both as a person and as a team member. It means you never settle for less than the whole thing. It means you can have it all and you can do it *as a team*. One of Dale Carnegie's greatest achievements was not that *he* became a millionaire, but that *thirty-nine of his best friends* became millionaires *with* him.

Unfolding your personal potential doesn't mean having to control everything. It doesn't mean denying anyone else the right to his or her own fulfillment. And it certainly doesn't mean clawing your way to the top. Unfolding means caring enough to overcome your *own* self-defeating behaviors so that you can contribute to *everyone's* success. You know you are "okay," so you mature your holodynes of denial, fear, and anger, and you move *beyond* your "fight-flight" games. You practice self-exploration and self-discipline, as well as self-assertion. And every time you do, you add to your team-building bank account. You earn the interest of self-confidence and genuine creativity.

Unfolding your personal potential means that, whatever you

do, you take the time to go to your place of peace, call up your Full Potential Self, and fully discuss the merits of what you want. Using your I.S.P., you learn to communicate clearly with your Full Potential Self. You *listen* to your "I"—carefully, fully, with your whole being. You *talk things out*. If you are not in complete agreement, you find out why. You get to the immature holodynes which are causing the trouble and you potentialize them. You align your mind, your activities, and your focus with your Full Potential Self and then you act accordingly. Everything in your body and mind comes into harmony with your focus.

These actions add to your "team power" bank account because self-directed change is the beginning of any team change. The first order of change occurs at the first order of organization, among the holodynes of your mind. So, carefully cultivate the garden of your mind. *Feed* your holodynes. *Challenge* them: read uplifting books, collect inspiring thoughts, have stimulating discussions, accept creative opportunities, do something spontaneous, accept difficult assignments, meditate deeply, keep a journal, do the "impossible." Think thoughts that others dare not think and dream dreams that others dare not dream. Keep your holodynes on their toes.

And keep the garden of your mind productive and weeded. This clears your holodynamic plane of any conflicting holodynes, strengthens the intensity and clarity of your focus, and broadens the range of patterns for the unfolding of your "I." You become an excellent team member.

3. GO FOR THE BEST DEAL POSSIBLE

The best deal possible is one in which *everyone* wins. Empowering a "win-win" relationship means that you keep your focus on the other person's fullest potential as well as on your own. This aligns your intents. It feeds the living dynamic between you. It keeps the deeper orders of your minds in harmony and it builds a reservoir of life energy, of emotional rapport, mutual trust, and genuine friendship. It lays the foundation for a bonded, effective, fully potentialized relationship.

This doesn't mean that you are "chained" to the other person. "Bonded" means "*inter*dependent," not "*de*pendent." It means you interact freely, along the lines of your *mutual* commitment. It means you don't have to manipulate the other person to get what you want, and you don't have to match the other person's images or expectations. Each of you sees the other as a "winner," each treats the other with the utmost respect, each supports the other even when not in the other's presence, and each protects the other's interest whenever necessary.

"Win-win" relationships go beyond competition. Life is abundant with opportunity and resources. There is enough for all of us. Through cooperation and communication, all of us can team up and dedicate ourselves to our common highest potential. In a "win-win" relationship, your success becomes my success, my joy, your joy. Your potential seeks release for me, my faith and action seek that same release for you. As "winners," we are self-motivated and self-disciplined. Gone are our childish needs to conform to set rules and to control others by constant supervision and impersonal memorandums. Our focus, deep bonding, open communication, and mutual commitment let us become completely aligned, and breathe life into our relationship.

But when, instead, you are raised on "win-lose" thinking, you interact in jarring, disconnected ways which create resentments and other negative feelings. Such feelings, even when not expressed, activate primitive holodynes, which then sabotage efficient and productive relationships. Your relationships become subject to illness, family pathologies, cynicism, emotional overreactions, and even fits of rage, until you attend to these troublemakers, track them, and mature them up, so that *everyone* wins.

The constant stress created by "win-lose" thinking breeds insecurity which, in turn, may cause you to burrow into your position of power, to collect credentials, to hoard possessions, to form secret affiliations, to get special concessions from rules and regulations, or to otherwise "fortify yourself" so that your security can no longer be threatened.

Like all polarizing mind games, the "win-lose" game can be played *both* ways: accomplished players can easily swing from "win-lose" to "lose-win" and back again. In "lose-win," you choose to lose and let the other guy win. Having fortified yourself with "win-lose" until your guilt, impersonal lifestyle, and empty relationships get to you, you drop into a suffering, "martyr-victim" role. You work as hard at *losing* as you ever did at winning. You "give up" control and "hang on the cross" for awhile, until you grow tired of all your "suffering." Then you swing back to "win-lose" and really "get" the other guy.

When you phase-space downdraft games, you can see how they operate at each stage of development. At Stage One, a typical downdraft game is "lose-lose"—where *everyone* loses. The Stage Two equivalent is the self-centered "win-lose" game, where you play to win, no matter what happens to the other guy.

If you "let" the other guy win ("lose-win"), you are at Stage Three—matching your images of suffering and solicitude. At Stage Four, you might legislate winning, as many special interest groups do, and even though no specific person loses, it's still a downdraft game because you use *external* means (the law, rules, or roles) to win without integrity. At Stage Five, you *believe* in "win-win" but you *don't follow* it. At Stage Six, you win the battle but lose the war. Whatever their differences, these games are all downdraft games.

The only way to win a downdraft game is to updraft it. So if you find yourself in a "lose-lose" game, redirect it—*updraft* it, or if you can't, then shift *out* of it and find another game which will updraft your energy. You can cut the cards but you don't have to play the hand.

If you are in a "win-lose" game, your crucial choice is whether or not to get yourself personally involved, whether or not to allow your "I" to *unfold* in this circumstance. If you do, just keep your focus upon your Full Potential Self and your actions will align with the best deal possible. If the *other* person wants to win by having you lose, shift out of the game by keeping your focus upon the *other's* Full Potential Self. It is literally impossible for anyone to play the "win-lose" game with you when your own energies are not resonating to the game. So focus upon what is really wanted by *everyone*, and the "win-win" philosophy will automatically take over. It only takes *one* person to shift the field. You must be clear. Your focus must be right on target. You must put yourself into the scene and keep your eye on the fullest potential for you and for the others, and then the best deal possible will potentialize.

When you negotiate such deals, call up the Full Potential Selves of *all* persons involved, sit them at your round table in your place of peace, and have a series of intuitive conversations with them *before* you begin the actual negotiations. In this way, your mind will resonate with good will, warmth, and genuine understanding about what is wanted on their part. You can listen with real intent, commit yourself to come to the best possible deal, wherein *everyone* will win, and know that your whole heart is in what you are saying.

With your intuitive guides keeping everything on track, you can explore alternatives more openly, more confidently, and more decisively. You can relate to your negotiating partners with greater consideration and empathy. You will show the natural courage and poise of a leader. People will look to you for solutions to their problems. Their trust and confidence in you will be well deserved.

4. ALIGN YOUR ACTIONS

Maturity means more than just producing results, no matter what the cost. So keep the total picture in mind at all times. Learn to see the end of your journey before you take the first step. Apprise each team member of his or her part in the production and align *all* your actions with the potential to be realized. In order to do this, you must have a clear blueprint, a clear plan of exactly what you want to accomplish, and then you must align your actions with your plan.

From such a blueprint of *where* you want to go, your plan of action—*how* to get there—naturally evolves. If you were building a house, the blueprint would give you a good idea of what your house would be like when you finished it. You and your team members would need to pour every bit of cement and drive every nail—align *all* your daily actions—with that blueprint in mind. Otherwise, it wouldn't take long for your efforts to come to nothing. With every step you take, you must keep your end in mind.

Now, systems grow the same way everything else grows—according to the six stages of development of the Implicate Order. So when you draw up your system's plan of action, be sure to keep all six stages in mind. At each stage of its development, your system must be carefully nurtured so that it will grow. Your system's physical needs must be met. Everyone involved in your system must feel that it's part of his or her real life purpose. You and your team members must unfold part of your Full Potential Selves into the system and align the power of your mature holodynes with the team effort. Your relationships must develop so that everyone feels good about working together, everyone contributes, and everyone's commitment is sufficient for what must be done. Your system—and what it produces—must fill a real need in the larger community. Then all actions will lead to synergy.

Synergy, that *extra* energy which comes from joining forces and working together, is the *natural* result of aligning your actions with your common intent. It is the Quantum Wave's response to the resonating frequency of your aligned, collective action. You can tell when team power is alive and well by the amount of synergy it creates in your system. When your team's collective thought processes are aligned with your common purpose, team members communicate, negotiate, and cooperate in updraft modes. They can phase-space any downdraft dynamic and align their actions so that *everyone* wins and synergy runs high. This is why tracking and potentializing are essential tools anywhere that teamwork is needed: they can *maximize* your team's synergy—maximize the amount of the Quantum Force coming through the wave in response to your aligned actions.

5. ADHERE TO PRINCIPLED PROCESSES

Because we can communicate instantly over the entire planet, because we are open to international commerce, and because we are naturally curious about one another, our world grows smaller every day, and more interdependent. We must be prepared to deal with different cultures, different ways of thinking and feeling, and different ways of doing things.

Even in our own culture, people who think differently are constantly being called upon to work together. This can strain relationships and cause deep disagreements. Some people value their differences and have learned to take advantage of them in the working world. Others resent and resist these differences. They don't know how to look *beneath* the differences to find their positive intent: they don't know how to phase-space.

Once you learn to phase-space, to look at differences from another space and see how these differences can be phased into working *together*, you are free to interrelate with almost anyone on the planet. You and your team members arrive at a principled perspective which promotes openness, cooperation, and complete alignment of all your actions. You come to care about one another, share with one another. You come to believe in fairness and to think only in terms of "win-win" relationships. You *"become"* your principles. You do this intuitively, deeply, and completely. Your partners know you are what you say you are, and you do what you say you will do. There is no question about it. You have a deep, abiding sense of integrity. You are attuned to oneness.

6. EXTEND YOURSELF

There is something fascinating about watching one of those time-lapse films of a flower opening into full blossom. Something in each of us identifies with nature opening into the fullness of her splendor. It is what we *want* to do the most, yet often do the least. It is our last, *least* attained stage of development. Most of us know that creativity can be a lasting condition, that going into projects with everything we've got can become a habit, or that loving can be an all-consuming passion for a lifetime. We've seen these truths demonstrated — in our lives, in other's lives, and throughout history. But still we choose to ignore them. We settle for *less* than we are.

Compare the flow of your life energy emanating from the Quantum Wave to a stream of water. Compare your "I" to a tap which turns this energy on or off, depending upon your choice. If you choose to updraft your thinking, you turn the tap on even more. If you choose to downdraft your thinking, you turn the tap off. The higher you progress in the stages of your development, the more life energy you control. When you reach Stage Six, you have aligned your physical, emotional, interpersonal, social, and principled holodynes and integrated them so that you can now extend yourself into ever-widening circles of influence in the world.

At each stage of your development, you create a different resonating frequency. Now, as you extend outward, all that you have done and all that you are becomes part of your total resonating energy field. If you have empowered your mind at each of the previous stages of development, but downdraft your thinking at this last stage, by *not* extending to your fullest capacity, you turn off your greatest possible life energy.

If you don't extend, you fall back. You procrastinate. You seek to excuse yourself. You detach and you become remote. You may feel you have done *enough* and so you elevate yourself above your fellowman. You may even become obsessed with your own glory, position, or power and, at the furthest extreme, become a tyrant.

To avoid getting caught in downdraft thinking, you must extend yourself to your fullest capacity. You must give yourself permission to live life fully. This is what is meant by the saying, "You keep what you share." When you receive from others, you learn and you grow, but as long as you only receive from others, you remain in the role of a child. When it is time for you to "come of age," your next step is to extend yourself, to give to others, to teach others what you have learned. So, as soon as you have learned the principles and processes of this book, teach someone else and you will learn them again in another way, a more advanced and extended way.

But extending yourself means more than just teaching. It means reaching deeply into your personal life and the lives of those around you. It means not only sharing your insights with others but sharing *yourself* with them, loving them, and enjoying life with them.

Filled with vitality, confident in your creativity, able to share your intimate nature with others, you connect with synergy. Your integrity inspires oneness. Your teammates can depend upon you because they know you and love you just the way you are. Differences become complementary. Each team member's strengths are viewed as an asset to the whole team. Like members of an orchestra, you know the score and play your parts in harmony with one another. You are a team.

9

Principled Well-being: Holodynamics and Harmony

Harmony is our natural state of being.

Harmony begins within. We resonate to the harmony of music, admire a team that makes the extra effort together, and we are inspired by the teamwork which goes into a successful space launch. When the cells of our body work in harmony, we experience health, and when all the holodynes of our mind work in harmony, we have peace of mind. Once a person learns to phase-space, use the Mind Model, track and potentialize those holodynes which create disharmony, anyone can create harmony within and resonate to the deepest natural beat of life. Peace is the by-product of a mature mind. It comes when our inner dialogues are working in harmony, when our holodynes have learned how to get along together, how to live life to the fullest. Then we have shifted into that frame of mind which *is* peace.

The kind of peace which *is* peace cannot be found by looking. That part of your mind which *seeks* peace is not the part which *knows* peace and *is* peace, because the holodynes which *seek* are not the same as the ones which *know* and *are* peace. Your Full Potential Self *knows* and *is* peace. This is why the power to create peace ultimately rests in your "I." Living life to the fullest means unfolding your Full Potential Self to *its* fullest. If you fail to do this, if you downdraft, you will create chaos.

Living a fully potentialized life creates an environment of peace. Peace is to your mind what beauty is to a flower and, like a flower, your life begins deep in the dirt. Born in trauma, you are forced to face non-peace from the very beginning. Your parents, siblings, and culture then teach you skills of survival and war. It is this experience with non-peace which eventually allows you to seek and understand peace. Peace exists as an enfolded potential woven into the fabric of your nature and hidden within the deeper orders of your mind. It is part of the Implicate Order of life.

After years of helping people work their way out of the chaos of their lives, I learned that misery, suffering, addiction, stress reactions, loneliness, depression, and every affliction the human mind can impose upon itself have a purpose. When I first began to phase-space such afflictions, underlying *principles* emerged. These principles, I found, had great power and, even though they were mostly subconscious, they gave structure to the whole flow of the thought streams of the mind. When I came to understand these principles better, therapy became easier.

Using the Mind Model, I began to observe these principles in action on the holodynamic plane within each of my patients. I learned that when they or their families were in chaos they were close to breaking through to a new order. If they updrafted into the new order, the result was a new depth of peace and harmony. They would weather many storms, experience all kinds of chaos and turbulence, and become, through it all, sensitized to their yearnings for some unfulfilled potential. The chaos itself would challenge them to unfold that potential. If they didn't rise to the challenge, the chaos would become stable and they would be locked into it. I learned that certain principles could be used to guide people through their chaos and into a new order of mind so they could once again be at peace or in harmony. These underlying principles would emerge at just the right time. Over the years I have collected these principles. They are the holodynamic principles.

The holodynamic principles evolved from two decades of phase-spacing what worked for people as they solved their problems. They began to emerge when I realized that personal troubles cause people to grow, to *potentialize*—that problems act as stimulants, invitations to live life more fully. Once a person understood that problems are invitations to grow, he or she could face them with an entirely new outlook. Rejection became an invitation to explore new bonding processes, anger became a sign of intense caring, depression a percolating potentiality, and death an invitation to intuitively connect with parallel dimensions.

I found that phase-spacing from a principled perspective immensely enhanced getting from point A to point B, whether it was helping someone in therapy, building a relationship, or making a business successful. It was a way to keep whole streams of thought within appropriate channels. If you wanted more money, the principles of abundance became guidelines by which you could manifest that abundance. Key among these was "I *am* abundance." Once you accepted this principle, you no longer depended upon "luck" or anyone else to provide you with abundance. The "I *am*" principle made clear your responsibility for setting up the resonating frequencies, aligning your Full Potential Self with your personal actions, relationships, and systems so that they could potentialize your abundance.

Such guiding principles became evident in all fields of human endeavor. Health, happiness, and success of any kind were guided by a deeper order, a principled dynamic. I discovered that principles have their own power. Love, for example, can be viewed as a living principle. It has power in people's lives. The same could be said of faith, trust, fairness, and equal rights. They are, in every sense, living entities within our collective mind. They are part of the fiber which builds our cooperative world of the future. Because this fiber is so important, and because I see our society as a living system evolving within the Implicate Order of growth, I would like to discuss the key principles by which I believe we will move from a society at war to a society at peace.

The journey from war to peace is a passage we all must make. If we fail individually, we are sure to fail collectively. If we fail collectively, we will surely destroy our civilization and end life on Earth. There is no more important quest for us than to learn to live life fully in a peaceful world. Each holodyne resonates with a frequency which ripples into the field of our collective consciousness. Therefore, we must all address our personal passage from war to peace. We can all potentialize together. In our business dealings, our competitions, and all our relationships, we must come to appreciate our life potential and see how it unfolds best in an environment of peace. Part of our potential for life requires that we establish in our *outer* world the peace we *are* in the deepest orders within orders of our minds.

Such peace lies enfolded within each mind and within our collective potential. By using the Mind Model and learning the skills of phase-spacing, tracking, and potentializing, you can discover the guiding principles which create harmony within your own mind. No matter how troubled your life has been, you can make peace a reality, by first bringing your conscious self in harmony with your Full Potential Self, bringing peace to your holodynamic plane, *being* at peace.

Once *you* are in harmony, you can stabilize peace in your relationships, on the job, and in the various organizations to which you belong. You can discover the underlying principles which guide all your efforts for peace. And knowing these principles, you can help potentialize peace in the world. These principles are the holodynamic principles.

If you want peace, inner peace or peace in this world, you must align yourself with the holodynamic principles. They are the invisible, underlying channels, the hidden streambeds which guide the flow of life energy. They can release humankind to reach its fullest potential. There is no more promising hope that we can turn back from our helter-skelter rush into total war than the knowledge that these principles are at work throughout the universe, that they will unfailingly bring us to our fullest potential and to peace if we but understand and live by them. They hold the keys to unlocking our individual and collective potential. They hold the promise that we can all live life to the fullest.

As I came to understand these truths, I condensed them into eight basic principles. From these, all dimensions of the Mind Model evolve, and the underlying structure of the holodynamic universe can be understood. There are others, but these eight are the most powerful:

1. The universe is holodynamic. All matter, energy, and intelligence—past, present, and future—are part of one, dynamic whole.

Once you realize that you are part of everything and everyone, the dynamics between you and every other person change. You realize that how you treat another is a reflection of how you treat yourself. All things become extensions of your own intelligence. You realize we are all part of the quantum field of reality in which each person functions as a major determinant of that reality.

From this perspective, your Full Potential Self may be compared to a shining light bulb inside a black box. On one side of the box is a hole. As the light from your Full Potential Self shines through the hole, it creates a quantum, multi-dimensional, holographic image, which becomes *you*. It takes time—nine months in the womb, and years of growth—but *you* gradually take shape, or unfold into physical reality. We are all like pictures on a movie screen—complex projections of what our Full Potential Selves want us to be in this plane.

Suppose, however, that on the other side of the box there were another hole. Your Full Potential Self could also be projecting itself through that hole into another world, a world *parallel* to this one. In fact, there could be endless holes in the box. So, as John Wheeler suggests in his many-world interpretation of quantum mechanics, we could be living an *infinite* number of lives *all at the same time*. So people who believe in past lives, Karma, or spiritual beings, people who believe they are someone else, or who have fantasy dreams, might be experiencing something real from a parallel world.

Once you understand that the *universe* is holodynamic, you can see that whether other worlds are real or imaginary is irrelevant because they are *all* holodynamic. They are all part of one, unified, cooperative force in action. Your Full Potential Self is able to connect with each dimension of its own reality. In therapy, this connection proved to be the single most powerful force for healing I ever experienced. Once my patients connected with their Full Potential Selves, they took charge of their own therapy process and success was theirs in a minimum of time. Their Full Potential Selves could take them back into life regression experiences which specifically pinpointed where their holodynes, the ones blocking their progress, were formed, and then guide them through the potentializing process.

To your Full Potential Self, your "I," all dimensions of the mind are holodynamic. Your "I" has the power to access and make available to you information regarding *any* dimension anywhere, anytime—past, present, or future. By implication, the past and future are going on *now* in some parallel world. Not only that, but every variation of your life, every decision you *didn't* make, every choice that went the *other* way, every job you *never* had is taking place in some other dimension. You are potentially experiencing everything possible in some world. You might even be your worst enemy in some world. Can you understand now why everything you ever believed is not true? The truth may be much grander than anything anyone has ever believed.

As your conscious mind gains in sensitivity and awareness, you are able to connect with all aspects of the holodynamic universe, whether in parallel worlds or within the deeper orders within orders of your mind. Your personal power grows far beyond anything you ever believed possible. You unfold your "I" in the now and experience the past and future at the same time— *holodynamically*.

2. The universe contains living thought-forms, called "holodynes," with the power to manifest reality in all dimensions.

Your state of being is controlled by the holodynes which are active on your holodynamic plane. When your holodynes are at peace, you are at peace. When your holodynes are at war, you are at war. You move from chaos into harmony by creating a new internal order within your mind: you potentialize all your holodynes and, accordingly, *yourself*.

Holodynes are the first order of organization within your mind. Understanding how to access and potentialize your holodynes gives you the necessary power to change your mind at its most primary level of organization, to clear its channels, to organize its holodynamic plane. It gives you the choice as to which holodynes you want to be active in your life, and thus another degree of freedom in determining your destiny.

Most people remain forever trapped in their primitive thinking because they cannot tap into the *source* of their troubles—the primitive holodynes which are programming their behavior. In therapy, education, business, and life, controlling the world of holodynes allows you to take charge of your life, and frees you to live life to its fullest.

3. The universe has an underlying, enfolded order—the Implicate Order.

All life is subject to boundary conditions. Every species lives within certain environmental, chemical, physical, and biological boundaries. Beyond those boundaries, life can no longer exist. There is order to all life. The source of this order is found within a *deeper* order, the Implicate Order, not readily apparent to our rational minds. As our technology improves and our thinking develops, this deeper order becomes more apparent. One way to identify this order is to phase-space growth itself.

All growth takes place according to a universal, built-in order. By consciously aligning the faculties of your mind and body with this order, you can accelerate your growth to its maximum potential. *Anyone* can do it. Those who do *not* align themselves with the Implicate Order of their own growth fail to grow. Those who *do*, find it such a powerful freeing dynamic that they experience quantum leaps in their growth.

4. Within the Implicate Order, the mind, holodynes, human beings, and all manifest reality follow the *same* six stages of development.

The fact that the Implicate Order had six stages which applied to *everything* astounded me. I found this was so by first carefully studying how humans developed through their various stages of growth, and then observing that these same stages applied to holodynes. That was a breakthrough for me. I worked for almost five years with schizophrenic families and observed their family systems grow through exactly the same stages. I knew then there must be some deeper, universal order.

I have phase-spaced individuals, groups, organizations, and systems, and observed, in *every one of them*, the same stages of growth. I have learned that all holodynes, people, relationships, and systems must progress through these same six stages of development to reach maturity—to potentialize. Once you understand this, you can learn to focus on *each* stage of development and systematically empower your own growth and your part in the growth of others.

5. Manifest reality has both a "particle" and a "wave" function, which the mind reflects through its rational and intuitive processes, respectively.

Being able to use both your rational and intuitive mind at the same time, increases your personal power dramatically. It is like changing from using only one leg to using two. It gives you the ability to intuitively sense the whole field (the "wave") of reality and, at the same time, to rationally deal with its parts (the "particles"). It gives you balance.

6. Change occurs holodynamically: to change any holodyne is to change the physics of the mind; to change the physics of the mind is to change the physics of the universe—past, present, and future.

Real accomplishment or change, real success in *any* endeavor, starts with your mind. And the first order of organization of your mind are holodynes. Learning to track and potentialize holodynes is fundamental to your success: when you change a holodyne, you change the resonating frequencies which emanate from your mind, and thus your life, the lives of others, and the universe itself.

7. Every human being has a primary, controlling holodyne, called the "I" or "Full Potential Self."

Your "I" or Full Potential Self orchestrates all the events in your life. Your "I" emerges in this plane and experiences life as you, the individual, live it. Your free choice determines the bulk of your experiences and creates feedback to your "I," which can change anything in the game of life within the boundary conditions of reality. By keeping your focus on your "I," by communicating regularly with it, and by following its advice, you can fully potentialize.

8. The holodynamics of the mind can be applied systematically to solve every problem of human experience.

Once you understand the holodynamics of your mind, you can tap directly into the Quantum Wave and utilize the Quantum Force. You can phase-space any problem to see its intent, find a mature way to reach that intent, and potentialize the solution. This frees you to live life fully.

These principles work. I invite you to explore them in your life. They complement the new quantum sciences, reflect the latest thinking in unified field theory, fluid dynamics, chaos theory, and systems analysis in transactional therapy. They provide new models for education, rehabilitation, and political reform. They will help you move from point A to point B. They will help you get what you want. They will help you live life to its fullest.

10

Universal Well-being:
Living Life to Its Fullest

You can phase-space the world—the whole dynamic of the planet. If you view some of its parts, you see chaos. When you look at the whole dynamic, you see a deep order of harmony— oneness. The more you live life to its fullest, the more oneness you experience.

As you experience this oneness, you will have both the rational and intuitive sense of it, see both its "particle" and its "wave" functions, and align yourself with both. Using Intuitive Sensory Perception in balance with rational analysis gives you the ability to experience the oneness, so as to give maximum meaning to it in your daily life.

Oneness is also created at each level of the Implicate Order of growth. As you updraft your thinking and act accordingly, your life energy goes into the world to strengthen, uplift, encourage, and support every life effort. This gives you a clearer view of how life supports you. When you encourage every healthy cause and represent abundance and vitality in all you do, you can sense how life gives you health and abundance in return.

As you choose to unfold your personal "I" and recognize your ability to impact the quantum field of the world itself, you become more conscious of your part in the whole world. You think in terms of world issues. You direct your effort toward world programs and you take interest in its past, present, and future. You can be more creative, more self-aware, and able to assert your Full Potential Self in a worldwide sense. You become seasoned at personal and collective commitment, active, aware, and sensitive about world needs. You can see the unfolding of the life potential of the whole planet.

You understand and experience the living principles of the holodynamic universe. You are able to respond to the life of the planet because you know it is holodynamic—*your* life is one with *all* life. You are able to attune yourself to every dimension of its reality because you know that reality is an extension of yourself. It is a quantum field. The universe, all matter, energy, and intelligence are part of one, dynamic whole. You use your loving, empowered, and knowing self to keep humanity and all of life evolving in a healthy way. You resonate, act, and impact the field so that all life can live fully, so that every life-form gets the best deal possible. You become the embodiment of oneness.

You can phase-space humankind, society in all its variations, and, using the Mind Model, examine its developmental stages. Societies are all going through their own stages of development and, put together, each contributes to the growth of humankind as a whole.

Some societies, for example, are very rational in their orientation, others are more intuitive. Some focus on material wealth, scientific discovery, or power. Others focus upon religion, gaining enlightenment, or living at peace. Sometimes these differences are embedded deeply in tradition and culture and cause continual conflict between cultures as the world evolves toward a universal society.

Phase-spacing helps pinpoint where such conflicts arise and what can be done about each one. Usually, the solution is found by focusing on the deeper order within the chaos. For example, in some societies, such as nomad hunting societies, unity is seen as necessary mostly for physical survival. First they hunt together, then they develop common defenses against predators—including other hunting societies. Next they hunt for females and develop a code of ethics which makes the female submit to the male. Male dominance is justified by the original premise—as necessary for physical survival. It is woven into the holodynes and resonates into the cultural field.

Physical survival becomes the controlling principle not only for male dominance over women but for all manner of personal and collective whims. It is used to create polarization and movement in funding defense budgets, in legislating special programs for people with special interests, and in justifying maintenance of industry harmful to society as a whole. In this regard, society is much like slime mold. Slime mold feeds on vegetation on the forest floor. Each individual slime mold cell grows and reproduces independently of the rest until the food supply runs low. Motivated by crisis, the cells all get together, form into one body which resembles a slug, and crawl over the forest floor until they locate a new area where there is sufficient food. Then the slug grows a long stalk, which develops a "head" on top. This head explodes and sprays individual slime mold cells out over the forest floor to begin the cycle once again.

When you phase-space the underlying principles which guide various societies, you see that physical survival is a strong motivation for social unity. It is built into the order of things. It is the first stage of development for a fully functioning society, but it can be downdrafted and used to justify almost anything. When you updraft it, physical survival becomes the effective use of all physical resources for the good of all.

Physical survival includes developing adequate systems for energy supply, food distribution, communication, transportation, sanitation, and health. We call a country "developed" when it has created these systems within its borders. But the world as a whole has only begun to create these networks *between* nations. When Buckminster Fuller phase-spaced the number of kilowatt hours available per person in various countries, he found some interesting comparisons. The energy available per person was directly correlated to infant deaths, educational standards, and health statistics. Fuller proposes an energy-saving worldwide network of power lines so that peak power times in cities could be coordinated throughout the world. This is already being done in Russia, which spans ten time zones, and also on the west coast of America. Power from electrical plants is transferred from low peak to high peak time zones. It has saved billions of dollars and kilowatt hours of energy and could be expanded into a network which could save countless lives and accelerate the development of less developed societies.

Physical development is only the first stage of social evolution. The second stage has to do with the emerging identity of the planet as holodynamic. World wars have been followed by a new order of cooperation, the United Nations, international trade groups, and regional alliances among nations. The Space Age has given us a clear picture of our planet without national boundaries. People are beginning to grasp the idea that our atmosphere and natural resources are limited and must be carefully protected and developed. Mass media, cultural exchange programs, and international information exchange have been encouraged by the high technology explosion. We have great potential, but as a world society, we are still in our youth. We are at the doorway of having unlimited power and energy, but we have no way of controlling who gets the power. It's like having a loaded gun in the playpen.

Sanity begins on the holodynamic plane. Each national leader must be surrounded by people who know how to address individual mania in an objective, effective manner. This is the only way to keep it from becoming collective mania. It takes people who know how to track and potentialize, who know how to use such things as the Mind Model, who can phase-space, and who understand the principles of holodynamics.

As a therapist studying chaotic families, I learned that chaos could be overcome by understanding the basics of holodynamics. My traditional, rational, linear training did not work. People did not overcome their problems until they addressed them holodynamically. Until they learned how to use their I.S.P. and their Full Potential Selves, how to potentialize each circumstance, they made little progress and therapy was long and cumbersome. The new science of quantum mechanics, the discovery of the Quantum Force, the Quantum Field, and holodynes, and the use of phase-spacing and the Mind Model, help solve individual, group, and systemic problems. These tools help identify the Implicate Order of growth for all aspects of society. They make it possible to focus and potentialize the world as a whole. They are part of the new order which is emerging from world chaos to bring us to oneness.

Oneness can only occur when everything is aligned—when every holodyne, every Full Potential Self, every person, every group, and every country is focused upon our fullest, collective potential. We must align ourselves so that everybody wins. This can only be done when we teach everyone how to use his or her I.S.P., to track, and to potentialize. The study of holodynamics is the key to oneness. It is the study of oneness. The flow of human evolution, the unfolding of life's fullest potential for the planet, can be enhanced by each individual effort when we are guided by holodynamic principles. We can extend ourselves, in tune with the holodynamic potential, reflecting the love we are, accessing the Quantum Force, aligned with our Full Potential Self, operating at full capacity, and move the world into a new stage of evolution.

A word of caution. We cannot confine oneness to our three-dimensional world. What the quantum thinkers are pointing out, and what my experience in therapy reinforces over and over, is that the quantum field is an *unmanifest* dimension. It is not obvious to the rational mind or the rational senses. It is only accessible intuitively. This is particularly evident on the holodynamic plane. Holodynes come in all shapes and sizes, some from worlds totally alien to ours. Intuitively, we know they come from parallel worlds. Some have influence in this world from one generation to the next. Others are invited by resonating frequencies which seem to create a field which allows them access. They can be "good" or "evil," angels or demons, mature or immature, effective or ineffective.

Holodynes can all be tracked and matured. They all respond to the same order of growth that we do. They are awaiting our loving interaction. They want to be freed from their spiritual prison and *potentialized*. They are continually confronting us, in our personalities, our organizations, and, as their influence impacts the world field, they play world war games. This is why it is necessary for us to begin to play world *peace* games. We can change the field.

Changing the field requires that we focus on what we want. That we learn to clearly identify the potential within each holodyne, person, relationship, system, principle, and within the whole. That we learn to reach in with our personal cup and collapse the wave of limitless potential into the one we want. Once we understand that change occurs holodynamically, we can see how changing even one holodyne changes the physics of the mind. It changes the resonating frequencies which emanate from the mind into the universe and therefore it impacts the universe.

From a quantum perspective, any change in the present has impact upon the past and future as well. Time is viewed as an oddity in space. We live in a "special" dimension which is limited to linear time. But the quantum dimension has no such limitation.

So any change, even in a single holodyne, has to meet the boundary conditions of the past and future before it can be manifest in the present. These boundary conditions are checked out by our Full Potential Selves, our "I's," and correlated with every other "I." In this way, collective oneness can be maintained at the most fundamental level within the quantum field. I have given examples of how we can go back through thousands of years to track holodynes and heal relationships, how we can *change* the field of the past. But we must do this *intuitively*. And we *can* do this because every "I" continually operates from a universal level and coordinates everything from a place of planning, which exists in the quantum field beyond the limits of time.

Once you are aligned with your Full Potential Self, you become a master at solving problems: you simply phase-space them, focus on solutions, and *potentialize*. As you practice tracking and potentializing, as you extend yourself and teach others to understand holodynamics, you become attuned to life at its fullest. Your love power and your sensitivity expand. You experience oneness and, in that oneness, you experience the ultimate unfolding of your Full Potential Self.

Appendix

Genogram Guide:
How to Make a Family Network Tree

The genogram allows you and your family members to see your family tree in a helpful, creative light. It is an excellent tool for phase-spacing your whole family system and, using the Mind Model as your guide, you can identify inherited family diseases, as well as critical emotional and relationship patterns, personality characteristics, and family beliefs which are passed on from generation to generation. This will help the next generation in overcoming family diseases, and all family members in handling family problems, improving family relationships, and cultivating family strengths.

When making your genogram, try to be as detailed and as comprehensive as possible. Include both negative and positive emotional patterns, all significant details of medical histories, and so on.

Also remember that, while climbing family trees can be enjoyable, it is often exhausting and sometimes threatening. Encourage all family members to contribute their best to the construction of the genogram. Usually, the more help you get, the better your genogram will be.

THE SEVEN DIMENSIONS OF YOUR GENOGRAM

Your genogram has seven basic parts, each of which will prove helpful in its own way. Try to complete as much of each part as you can.

1. The Family Tree:

Begin with a diagram of your family tree. Use a circle for each woman and a square for each man. Draw your own position, then your spouse's and your children's, as shown in Chart 2 on page 208. This is called your "nuclear family," and it is the center or "trunk" of your family tree.

Now extend your diagram to include the various branches of your family tree. Keep expanding until you have your parents and all their children, your grandparents and their children, and, if possible, your great-grandparents and as many of their children as you can get onto the chart. Don't worry if you can't complete all the details. Just draw circles and squares for as many family members as you can.

Next, number your circles and squares and fill in first names and ages, again as in Chart 2. In this way, you will be able to refer to each person by number, name, or both. And, finally, add the dates of each marriage (M) and divorce (Div).

2. Medical Histories:

Tracing back the histories of family diseases and ailments can be very helpful if you know what to look for. Diabetes, alcoholism, disorders of the heart, pancreas, and liver are sometimes transmitted genetically. Arthritis, multiple sclerosis, stress reactions, and emotional disorders, such as obsessions, compulsions, aversions, and excessive guilt or sensitivity, are likely to be more significant than broken legs (unless, of course, your family has a *preponderance* of broken legs!). The idea is to keep an eye out for *recurring* diseases, conditions, or ailments in your family tree.

3. Emotional Patterns:

Look at how each person in your family feels about himself or herself, about others, about life. Some family members may be open, accepting, cheerful, easygoing, or optimistic. Others may suffer from depression, phobias of various kinds, severe temper, spitefulness, jealousy, or negativism. You can usually spot these patterns if you ask questions like "What are the first five words you can think of which best describe Grandpa?" Then compare how you see Grandpa with how others see him.

One family member might say, "Grandpa was grumpy 90% of the time," and another might chime in, "Yeah, and he was miserable the other 10%!" Knowing that Grandpa was always grumpy can help children realize how Dad developed some of his negative emotional habits. It can also help the present generation overcome these "inherited" patterns.

4. Relationship Dynamics:

Look at how the members of your family relate to one another. You might, for example, ask, "What kind of relationship did Mom and Dad have?" "How did Grandma handle Grandpa's grumpiness?" See whether family relationships are open or closed, blaming or exploratory, manipulative or negotiating. Discover how your family members handle crisis, who wields the power, who makes the big decisions, and who makes the small.

Label each relationship between family members, or between groups of family members, with the quality that characterizes it (distant, hostile, close), and relationships which stand out as special with capital letters (A, B, C), so you can describe them more fully on a separate page.

5. The Family System:

You will also find it helpful to look at how the various parts of your family system work together, or fail to. Are there any coalitions (special groups that keep together and keep others out), or special roles assigned to certain family members or parts of the family? Are there any fractures (divorces, separations, feuds), black sheep, or "problem" people? Can you see how the way your family system works gets passed on from one generation to the next? You may wish to use colored pens to circle special parts of the family network, so you can describe them in detail according to their color.

6. Family Beliefs:

Family members pass on to you their beliefs about every phase of family life: how to raise children, how to deal with adolescents, when and whom to marry, how many children to have, how to earn a living, what kind of work is best, how to measure success, how to handle crisis, loss, trauma, and tragedy, how to grow old, how to face death.

Pay special attention to your family beliefs: they are most likely what *you* believe, consciously or subconsciously, about how to survive and how to live. If they are immature, fractured, or dysfunctional, they can limit your thinking, block your growth, and keep you from fulfilling your potential. Exploring such beliefs is the beginning of aligning them with your fullest potential.

7. Society and Your Family:

Finally, step back and look at the way your family sees itself, as a unit, in the larger context of society. How does your family, as a whole, represent itself to society? With what other systems does your family identify? And how does society generally respond to your family?

In putting together your genogram, try to answer the following questions as accurately and completely as you can:

GENOGRAM QUESTIONS

1. What major illnesses (physical) are there in your family?
2. What emotional illnesses? (alcoholism, drugs, mental illnesses)
3. What deaths, and what are the causes of death?
4. What divorces or separations, affairs, secret involvements?
5. How would you best describe the personality of each family member?
6. How do family members express love and affection? How do they know when love and affection are expressed?
7. How do family members argue? How do they express anger? How do you know when they are angry?
8. Who was/is an extrovert, an introvert, non-verbal?
9. Who was/is the major provider, the main nurturer?
10. What alliances, coalitions, and subsystems are there in your family? What are their rules and boundaries?
11. What are your family's myths? What are its secrets?
12. How do family members communicate? (words, gestures, expressions, body language)
13. What are your primary values? What are your family's?
14. How is masculinity/femininity expressed in your family?
15. What are your family's "do's" and "don'ts," "shoulds" and "shouldn'ts"?
16. How are feelings acknowledged, communicated, or avoided in your family?
17. How are decisions made in your family? Who makes them? Who is involved?
18. How do family members behave and relate in public, as compared to private (at home)?

CHART 2: SAMPLE GENOGRAM

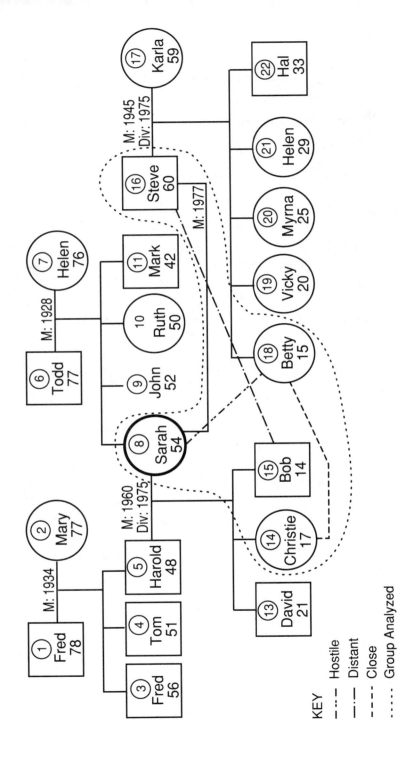

KEY

––– Hostile

–·– Distant

--- Close

····· Group Analyzed

For further information on how to use the genogram to create positive growth within your family system, you may wish to refer to the author's article, "Family Network Systems in Trans-generational Psychotherapy: The Theory, Advantages, and Expanded Applications of the Genogram," *Family Therapy* 10:3 (1983), available from Libra Publishers, Inc., 391 Willets Road, Roslyn Heights, NY 11577.

Suggested Reading

Backster, Cleve. "Biocommunications Capability: Human Donors and In Vitro Leukocytes." *International Journal of Biosocial Research*. Tacoma, WA, 1985.

Becker, Robert O., and Gary Selden. *The Body Electric: Electromagnetism and the Foundation of Life.* New York: William Morrow, 1985.

Bohm, David, and F. David Peat. *Science, Order and Creativity.* New York: Bantam Books, 1987.

Bradshaw, John. *Bradshaw On: The Family.* Deerfield Beach, FL: Health Communications, 1988.

Briggs, John P., and F. David Peat. *Looking Glass Universe: The Emerging Science of Wholeness.* New York: Simon & Schuster, 1984.

Eisler, Riane. *The Chalice & The Blade.* San Francisco: Harper & Row Publishers, 1987.

Gleick, James. *Chaos: Making a New Science.* New York: Viking Penguin, 1987.

Kramer, Jeanette R. *Family Interfaces: Transgenerational Patterns.* New York: Brunner/Mazel, 1985.

Nordenstrom, Bjorn. "Biologically Closed Electrical Circuits." *Discover Magazine* (April 1986).

Peck, M. Scott. *People of the Lie.* New York: Simon & Schuster, 1983.

Rossi, Ernest L. *The Psychobiology of Mind-Body Healing.* New York: W. W. Norton & Company, 1986.

Salk, Jonas. *Anatomy of Reality.* New York: Columbia University Press, 1983.

Sheldrake, Rupert. *A New Science of Life.* Los Angeles: J. P. Tarcher, 1981.

Small, Jacquelyn. *Transformers: The Therapists of the Future.* Marina Del Ray, CA: Devorss & Company, 1982.

Stone, Hal, and Sidra Winkelman. *Embracing Our Selves.* Marina Del Ray, CA: Devorss & Company, 1985.

Thomas, Lewis. *The Lives of a Cell: Notes of a Biology Watcher.* New York: Bantam Books, 1974.

Wilber, Ken, ed. *The Holographic Paradigm and Other Paradoxes.* Boston: Shambhala Publications, 1982.

Wolf, Fred Alan. *Parallel Universes.* New York: Simon & Schuster, 1989.

Glossary

Access To gain entry into, uncover, or become aware of, as when the conscious mind becomes aware of, or uncovers, **holodynes**, or "tunes in" to any of the dimensions of the **holodynamic plane**.

Boundary Conditions The requirements of the past, present, and future which determine whether a specific potential can be realized. When the mind focuses upon a **Quantum Wave** potential, its focus creates an echo wave, which goes back into the past, out into the present, and forward into the future, and which checks the conditions necessary for alignment with all levels of the **Implicate Order**, from physical to universal. Once these boundary conditions are met, the echo wave feeds back to the **Quantum Wave**, which then collapses into manifest ("particle") reality.

Causal Potency The power to cause, affect, or create.

Comfort Zone As represented in the **Mind Model**, the area of overlap between **family** and **cultural beliefs** of the **mind**, where beliefs are sufficiently comfortable, stable, and consistent to be held as "true."

Consciousness   Awareness; especially, that awareness of thought, feeling, and volition which arises when the rational and intuitive processes of the mind are integrated and in balance.

Cultural Beliefs   Traditions, customs, habits, attitudes, opinions, and persuasions held to be of collective value by non-family groups of a culture; those beliefs as experienced by an individual, stored in the **holodynes** of the mind, and occupying a designated area in the **Mind Model**.

Family Beliefs   Traditions, customs, habits, attitudes, opinions, and persuasions held to be of collective value by a family or any group of people living together in intimate relationship; those beliefs as experienced by an individual, stored in the **holodynes** of the mind, and occupying a designated area in the **Mind Model**.

Full Potential Self   The enfolded fullest capacity of a person; his or her primary **holodyne**, the "I," which influences and shapes experience through all stages of development, at all levels of the **Implicate Order**, and throughout all dimensions of the mind.

Holodynamic   (From "holo," meaning "whole," and "dynamic," meaning "force in action or motion.") Having to do with the universe as a projection of the **Quantum Force** through **holodynes** interacting at all levels of the **Implicate Order** to create physical matter, life on earth, and human experience within one, dynamic whole.

Holodynamic Plane   The plane of the mind where all thought patterns can be seen as part of the whole; in the **Mind Model**, the diagonal plane between the rational and intuitive portions of the mind, where activated **holodynes** operate and interact (bond); by extension, the plane of all life action in the **holodynamic** universe.

Holodyne   The fundamental unit of the **holodynamic** universe; the first order of organization on the manifest plane; a multi-dimensional holographic unit of memory storage. Holodynes are thought-forms with **causal potency**, which behave like living entities within the mind, creating, directing, shaping, and influencing (as **subtle attractors** and **selectors**) its thought streams; on the **holodynamic plane**, they encode the resonating frequencies of the **Quantum Wave** as it passes through them to all levels of the **Implicate Order**. Holodynes are responsible for ego-states, engrams, frames of thought, inner dialogues, personality characteristics, and behavior patterns; they are formed from sensory input, modeling, experience, imprinting, genetic inheritance, imagination, and **parallel worlds**.

"I"   See **Full Potential Self**.

Implicate Order   The enfolded, underlying order of the universe, out of which all ordinary experience is manifested. The Implicate Order consists of six interacting levels, from physical to universal, which correspond to the six stages of development.

Interest Wave   The ebb and flow of concern, focus, attention, passion, or curiosity, whether conscious or unconscious; within the mind, it is shaped and controlled by **holodynes** on the **holodynamic plane**. The interest wave has dimensions of intensity and frequency and is depicted topologically as going through the center of the **Mind Model**.

Intuitive Sensory Perception (I.S.P.)   The use of the senses by the intuitive mind. Physical reality appears to have both "particle" and "wave" functions; human sensory perception has a corresponding dualism. "Particle" functions are perceived by the rational senses and processed mainly in the left hemisphere of the brain by rational thinking, which is linear, analytical, logical, and technical. "Wave" functions are perceived by the intuitive senses and processed mainly in the right hemisphere of the brain by intuitive thinking, which is non-linear, creative, holistic, and artistic. Conscious use of I.S.P. allows the mind to directly **access holodynes** and the "orders within orders" of its thought streams and to intervene in their dynamics, as well as to integrate mental processes at levels of complexity beyond the realm of ordinary rational thought.

Manifest Plane   The world as we normally view it; that which is evident to, and perceived by, the rational senses, usually limited to the three dimensions of length, width, and height over time.

Mind   Intelligence as it encompasses all mental activity, whether rational or intuitive, manifest or unmanifest, quantum or physical, individual or collective.

Mind Model   A topological model representing the various aspects of the mind (**family** and **cultural beliefs**, rational and intuitive thinking, **holodynes** on the **holodynamic plane**, and the stages of development) and how these interface with the **Quantum Wave**.

Parallel Worlds   Worlds which exist in planes parallel to the **manifest plane** of our world. Such worlds are thought to communicate across a time continuum, wherein past, present, and future are one.

Phase-Spacing   The process of viewing any dynamic in terms of its "space" on the **Mind Model**. Phase-spacing allows us to consciously step outside of any problem and to view it from the perspective of growth and choice, as related to **family** and **cultural beliefs**, rational and intuitive thinking, **holodynes** on the **holodynamic plane**, and the stages of development—and to how these interact with one another.

Potentializing   The realizing of any potential by focusing upon it and by aligning it with the **holodynes**, relationships, systems, and principles involved, and with the universe as a whole. Potentializing occurs naturally whenever **boundary conditions** are met and the **Quantum Wave** collapses into manifest ("particle") reality.

Quantum Force   A universal force pervading the fabric of space and time. The Quantum Force is thought to be the source of all energy on the **manifest plane** and so powerful that a single cubic centimeter of space contains more energy than all known matter in the universe.

Quantum Thinking Thinking which combines rational ("particle") and intuitive ("wave") processes in a balanced, synergistic way and which allows the mind to grasp events, changes, and connections inaccessible or incomprehensible to either process of the mind used separately. Quantum thinking **accesses** the quantum dimension of the **Implicate Order** that underlies all manifest reality.

Quantum Wave A universal wave containing all possibilities on the unmanifest plane for any circumstance; also called the "Offer Wave." From the rational ("particle") perspective, the Quantum Wave is composed of an infinite number of "quanta," or discrete "bits and pieces," which together make up the complete set of possibilities for any given circumstance. From the intuitive ("wave") perspective, it is a field in which all possibilities exist without distinction until a mind focuses upon the field with specific intent.

Resonating Energy Field A field of encoded resonating frequencies, emanating from a source with **causal potency**, e.g., an "I" or a **holodyne**, and creating a growing environment in which matter or living entities are formed, take shape, or grow, as for instance, crystals, molecules, cells, plants, animals, humans, and social systems.

Round Table In **tracking**, a place within the **mind** where mature **holodynes** can meet and communicate with one another, with the "I," and with the conscious self, so that principled and peaceful processes can be incorporated into an individual's life.

Subtle Attractor The underlying and potentially powerful influence of any intent, whether conscious or unconscious, which attracts things, persons, relationships, systems, or principles aligned with it.

Subtle Energy Senses A biological sensory system, distinct from the normal five senses, extending throughout the body, wherein each cell acts as both receiver and transmitter of subtle energies.

The system is governed by the brain's limbic center, which processes sensory information and determines the body's responses through the hypothalamus and pituitary gland.

Topology   A branch of mathematics which deals with changing the shape of objects without changing their function; as applied to the **Mind Model**, a diagrammatic representation of aspects of the mind which does not change their function.

Tracking   The process by which a person, using **I.S.P.**, **accesses** a specific **holodyne**, befriends it, learns its positive intent, transforms it into its mature image (**potentializes** it), and commits it to a systematic, principled, universal relationship.